BELIEF AND PROBABILITY

# SYNTHESE LIBRARY

MONOGRAPHS ON EPISTEMOLOGY,
LOGIC, METHODOLOGY, PHILOSOPHY OF SCIENCE,
SOCIOLOGY OF SCIENCE AND OF KNOWLEDGE,
AND ON THE MATHEMATICAL METHODS OF
SOCIAL AND BEHAVIORAL SCIENCES



VOLUME 104

JOHN M. VICKERS
*Claremont Graduate School*

# BELIEF AND PROBABILITY

D. REIDEL PUBLISHING COMPANY
DORDRECHT-HOLLAND/BOSTON-U.S.A.

Library of Congress Cataloging in Publication Data

Vickers, John M
Belief and probability.

(Synthese library; v. 104)
Bibliography: p.
Includes index.
1. Judgment (Logic) 2. Belief and doubt. 3. Probabilities.
I. Title.
BC181.V5 121'.6 76–55341
ISBN 90–277–0744–8

---

Published by D. Reidel Publishing Company,
P.O. Box 17, Dordrecht, Holland

Sold and distributed in the U.S.A., Canada, and Mexico
by D. Reidel Publishing Company, Inc.
Lincoln Building, 160 Old Derby Street, Hingham,
Mass. 02043, U.S.A.



Printed in The Netherlands

# TABLE OF CONTENTS

# PREFACE

## 1. A WORD ABOUT PRESUPPOSITIONS

This book is addressed to philosophers, and not necessarily to those philosophers whose interests and competence are largely mathematical or logical in the formal sense. It deals for the most part with problems in the theory of partial judgment. These problems are naturally formulated in numerical and logical terms, and it is often not easy to formulate them precisely otherwise. Indeed, the involvement of arithmetical and logical concepts seems essential to the philosophies of mind and action at just the point where they become concerned with partial judgment and belief.

I have tried throughout to use no mathematics that is not quite elementary, for the most part no more than ordinary arithmetic and algebra. There is some rudimentary and philosophically important employment of limits, but no use is made of integrals or differentials. Mathematical induction is rarely and inessentially employed in the text, but is more frequent and important in the appendix on set theory and Boolean algebra.

As far as logic is concerned, the book assumes a fair acquaintance with predicate logic and its techniques. The concepts of compactness and maximal consistency turn out to have important employment, which I have tried to keep self-contained, so that extensive knowledge of metalogical topics is not assumed. In a word, the book presupposes no more logical facility than is customary among working philosophers and graduate students, though it may call for unaccustomed vigor in its application.

Two appendices are included; on set theory, including Boolean algebras, and on the rudiments of the theory of measures. These pretend to be not general introductions to their subjects, but may serve to introduce the unfamiliar reader to the concepts used in the book, or to renew a lapsed acquaintance. In either case it is recommended that they be worked through rather than merely read.

The philosophical presuppositions are not as easily specified. Most important, I should say, is an interest in problems about the nature of judgment and the relations between logic and belief. Certain works of Hume, Carnap, Ramsey, Quine and Peirce come to mind, but I hope that the text does not presume them.

## 2. A REMARK ABOUT SOURCES

There are a few words in the introduction about relations to views of Quine and Davidson. Donald Davidson is one of those to whom I am most indebted, and much of what is done in this book consists in working out or modifying things he has said. It is hard to document this debt, because most of our intercourse has been in his lectures and discussions at Stanford, and his published work has not attended to these matters in a way which would permit their isolation in reference. One source can be mentioned explicitly and clearly, and that is Davidson's question, "How could partial belief fail to be probabilistic?" Much of what I have to say is by way of trying to answer that.

Other debts which defy precise description are to Patrick Suppes and Richard Jeffrey. I owe much to Jeffrey's published work, in particular to *The Logic of Decision*, but I must also thank him for the unpublished development of his views on the principle of indifference, and in general for the approach to probability theory by way of belief and action. Suppes has encouraged me for some time to work on the problem of the relation of logic and probability, and much of the development of that relation has been with his counsel and assistance.

Another major source which is pervasive is the work of Rudolf Carnap. I have tried to say in a few places how what is there relates to Carnap's work, but there are many connections which go unremarked. They will be obvious to the student of Carnap's writings, but others should beware of attributing to me what may be found in *The Logical Foundations of Probability* or *The Aim of Inductive Logic*.

The classical sources, which should also include Carnap were it not for the relative recency of his work, may less unjustly go unremarked. They are mentioned at several places in the text.

CHAPTER 0

# INTRODUCTION

## 0.1. Logic and probability

The subject of this monograph is the relation of logic and probability. That is a large subject, and the work does not approach comprehending it. It is rather the attempt to develop a particular point of view about logic and probability, which is, roughly put, that the function of logic in the theory of probability is to specify the equivalence classes of sentences or propositions, such that members of the same class have always the same probability. That is the view, and a lot of the work of the book consists in finding out what the limits of this function are; just how much effect logic can have on probability, and to just what extent probability may be assumed to reach conclusions which are independent of assumptions about the logical relations among its objects.

One of the consequences of the investigation is the formulation of a condition, the *simple sum condition*, which is equivalent to the laws of probability and which makes the function of logic more apparent. This condition may be of independent interest, because it employs and gives some meaning to probability sums the totals of which exceed unity: It is usual in probability and statistics to sum the probabilities of alternatives which are known or assumed to be incompatible, thus assuring that the sum will not exceed unity. Little meaning could be attached to greater probabilities, since there could, by the assumptions of the theory, be no proposition or event, the probability of which exceeded one. The meaning which is attached to probability sums which exceed one by the sum condition, (they may even be infinite), is that the sum of a probability measure over a set of propositions must always (i) exceed the cardinality of some logically closed subset, and (ii) be exceeded by the cardinality of some consistent subset. The sum of a probability must always lie between the least and the greatest numbers of propositions that one can consistently assert, without asserting any others. In the case of a finite set of pairwise logically incompatible propositions, this truth is a consequence of the constraint of additivity as usually formulated, and is apparent upon brief reflection: In this case, the sum of a probability over the set is just the probability of the disjunction of its members. If the set is also logically exhaustive, then exactly one of its members both must and may be consistently asserted, and that is also the probability of the disjunction. If the set is not logically exhaustive, then it is consistent to assert zero or one of its members, and the probability of the disjunction must lie between these limits.

The sum condition, however, holds also when the sets in question are denumerably infinite, and also when their members are not pairwise logically incompatible. Any amount of redundancy is permitted.

The question of the relation of logic and probability is also a question of the relation of logic and judgment. At least it is so if judgments of less than full strength obey the laws of probability. The assumption that this is so is justified both from the point of view of the psychology of judgment and from that of the theory of probability. The latter justification does not, I believe, depend upon a subjectivistic interpretation of probability. Subjectivistic interpretations of probability may be distinguished from objectivistic interpretations in something like the following way.

Objectivistic interpretations differ from subjectivistic interpretations in their relations to thought or judgment. Objectivistic interpretations define probability with no reference to thought, and the relation to thought or judgment is just that of an object to the thought or judgment of it. It is, in the technical phenomenological sense, *noematic*. Subjectivistic interpretations, on the other hand, define probability in terms of characteristics of thought or judgment, such as strength or liveliness, or in terms of the actions to which those thoughts or judgments lead.

That objectivistic probability is defined without reference to thought does not mean, of course, that it is or seeks to be ignorant of the nature of probabilistic judgment. It may be considered, and frequently is so by its proponents, to be an account of probabilistic judgment, according to which, for example, probabilistic judgments are simply judgments with probabilistic objects, and the mark of partial judgment will be in the partiality or probabilistic character of its object. On this account, the relations among probability judgments will be relations among their objects, and the question, for example, whether two probability judgments are compatible, will be the question whether the propositions judged are, in company with the laws of probability, compatible. In this way judgments of less than full strength, that is to say, judgments with probabilistic objects, will conform to the laws of probability.

In any event, then, and on any account of probability, to ask about logic and probability may involve one in questions about logic and judgment. In particular, in questions of the invariance or transparency of judgment for logically equivalent objects, of whether it is possible to judge such objects to different degrees, or to assign them different probabilities. The theory of probability as usually formulated does not address this question. In the customary association a proposition is thought of as a set of possible worlds. All necessary propositions are identified with the universal set, all impossible propositions with the null set, and logically equivalent propositions are identified with the same set. The disjunction of propositions is associated with the union of the sets associated with the disjuncts, and the conjunction with the intersection of the sets associated with the conjuncts. If one starts with a collection of propositions which is closed under the formation of truth-functions, then the resulting collection of sets of possible worlds is a Boolean algebra, and a probability defined for propositions may be understood as determined by a probability on the associated algebra. It is consequent upon this development that all necessary propositions are assigned one, all impossible propositions assigned zero, and logically equivalent propositions assigned the same probability. Thus the relation of logic and probability is assumed rather than investigated by the scheme.

## 0.2. Judgment

Belief is, in some essential part, a mentalistic concept. What a man's beliefs are depends in some way upon his awareness. The limit of involvement with awareness is with the sorts of belief called *judgment* and *assertion*, in the psychological meanings of these terms. In this sense every judgment and every assertion is the judgment or the assertion of a proposition at a time by a subject, who is aware, at least, of what he is judging or asserting.

Assertion is closely tied to language. In the clearest case a man speaks or writes to make an assertion, and the act of asserting is just that speaking or writing. Of course, not everything spoken or written is asserted; the speaker must know what it is that he asserts, and in the standard cases he intends by his assertion to reveal his conviction in the proposition asserted. The non-standard cases are important and difficult. Whether a man must believe what he asserts, whether lies are assertions, and whether one may in speaking or writing a sentence assert a proposition other than that expressed by the sentence, are difficult questions, and we are not yet clear on the import of answering them one way or the other.

There may be conventions which enable assertion without actual speech or writing. Signals, for example. But these instances do not loosen the bonds of assertion and language, for what is asserted is always capable of expression in language, and if not actually formulated in the conventions, is usually found in some essential subjunctive by which those conventions are tied to the act.

Judgment is very much like private assertion, if that is possible. One may make a judgment without asserting it, and then assert what he has judged later, or perhaps not at all. There are also cases in which assertion and judgment are simultaneous. probably not even distinguishable. Whether it is possible to make an assertion without a simultaneous judgment is not clear, but it is in any case possible to judge without asserting, and, indeed, without any outward declaration or indication that or of what a judgment is made.

The question of the logical relations among propositions about judgments is in large part the question what implications hold among propositions of the form

*X* judges that ____

Of this sort, for example, is the question of whether a man may judge the conjunction of propositions and yet fail to judge both propositions individually. Our inclinations here are that implication among propositions about judgments should be weak, and that few inferences should be licensed. We think that whatever a man judges must be before the mind, so while we tend to affirm the inference from

*X* judges that *P* and *Q*.

to

*X* judges that *P*.

we do not allow that from the latter to

*X* judges that *P* or *Q*.

nor do we easily admit that from

*X* judges that everything is *F*.

to

*X* judges that *a* is *F*.

Although the inference depending upon conversion, from

> *X* judges that *a* is *F*.

to

> *X* judges that something is *F*.

seems more binding. Negation internal to judgment is also puzzling, especially when it is mixed with quantifiers, and the best general remark about the logic of judgment seems to be that we have no satisfactory general account of it, even when we are quite clear about the logical relations among the propositions judged.

Many of the problems in the way of a theory of the logic of judgment stem from the difficulty of giving criteria for deciding when a man has or has not made a judgment. We all know what it is to make a judgment, but the phenomenology of it is elusive, and Hume's eloquent struggles to describe the conscious characteristics seem not to have been improved upon. Judgments need not be asserted, nor must they be otherwise manifested in the public conduct of the judger. We become sometimes so exasperated with the inadequacy of objective criteria that we are tempted to deny the meaningfulness of the concept, to deny that we can mean any more by saying that a man makes a judgment than that he would assert it if asked under the appropriate conditions, that he acts in accordance with it, and so on. As severe as this temptation is, it should be resisted, for to yield to it is to purge our experience of one of its most important elements, one which is quite subjective and at the same time cognitive.

## 0.3. Belief and Action

Judgment is the mentalistic limit of the concept of belief. It has only partial and inconstant connections with the agent's conduct. We have also a concept of belief of a more enduring sort. A man's beliefs, in this sense, remain his while he sleeps or is otherwise unaware of them, and, indeed, many of one's beliefs, among them many held with great security and endurance, are rarely, perhaps never, brought to consciousness. Now, a man may have beliefs of this sort which are never manifested in his conduct, if, for example, no opportunity for such manifestation arises. We are sometimes surprised, upon the occasion, by our own beliefs which we had not previously recognized, and we need only suppose that such an occasion had never been realized, to see the possibility that the belief had never been manifested. The view according to which the belief comes into existence only upon the occasion of its manifestation in consciousness or conduct will not adequately cover the facts. It would, for one thing, make it impossible to account for action by finding its origins in the beliefs of the agent, and it would make no sense of inferences made in conformity to a general belief which is not itself found in consciousness. We should thus abandon one of the main ways in which actions or inferences of the same agent at different times are related and explained.

So belief is not to be understood directly in terms of the agent's conduct, but when contrasted with judgment by its endurance, dispositionality, and casual relation with the agent's awareness, it does seem to be more closely related to action, and that in a different way, than is judgment. A man's beliefs are more public and objective than are his judgments. Since he may be surprised by his beliefs, he may be mistaken about them, both in the sense of judging falsely that he does or does not believe something, and in the sense of having incorrect beliefs about what he believes. We are much readier to contest a man's assertions about his beliefs than we are to contest what he says about his judgments. We do this on the basis of his conduct. In some cases, when a man asserts himself to believe what we are convinced that he does not believe, we think that he is being hypocritical, that is, that his assertion is contrary to his own conviction, but we frequently allow, and occasionally insist, that he is himself deceived.

When we attribute a belief to a man we say something about how he does or would act in certain situations. That is not all that we say, to be sure, but if our assertions about what a man believed had no implications for his ways of acting, then his actions could not influence our views about his beliefs as they do. In general we seem to take a man's beliefs to describe the world as he views it, and we try to read back from his actions, and assumptions about his desires, to reach conclusions about the nature of that view. In order for this process to function there must be some substances and events common to the world as we view it and as he views it. That is to say, that we must assume at least some of his beliefs to be true. In order to argue that he has certain beliefs, we must assume that he has others, some of which are assumed to be true.

Any attempt on my part to contemplate the totality of my beliefs encounters the difficulty that the contemplation of that totality, which may take the form of a belief, is not itself a part of the totality. That is an important feature of intentionality and it may be the ground for diagonal argument, but it need not be mysterious, and it should not occasion mystifying. On the other hand, this characteristic does mark an important distinction of objectivity from subjectivity which we ought not ignore: There is no difficulty of this sort, whether it is impossible for other reasons is not at issue, in the way of my contemplation of someone else's beliefs. That is because my beliefs about his beliefs are not *his* beliefs, and hence their exclusion from the totality does not prevent its completion. The question naturally arises what does my view of his beliefs have to do with his view of his beliefs, and should either of these views be taken as a standard which the other tries to approximate.

The concept of judgment is distinguished from that of belief in at least the following way: We give full priority to the agent's view of his judgments. He knows best what his judgments are. (Of course he may not know best whether they are true.) But when it is a question of what a man believes, we are more willing to allow for error on the part of the agent in order to improve the objective correlation of his beliefs with his action. The ground for this is that the agent has an authority in the regard of his own awareness which he does not enjoy with respect to his actions. This is not to say that no distinction can or ought be made between the agent's and the observer's view of an action, it is just to insist that the agent's view is corrigible, since his action is public, and that an observer may know something about an action which the agent himself does not know. To the extent that belief has implications for action, mistakes about the nature of actions may entail mistakes about the nature of beliefs. Thus the agent himself may be mistaken about what he believes, but not about what he judges.

A man's beliefs are not causes of his actions. There are two strong theoretical reasons for this. First, his beliefs are characters or dispositions, not episodes or events, and second, what his beliefs are is in some part logically or essentially dependent upon what his actions are. Thus we discover his beliefs by observing his actions, not by means of an inference from effect to cause, but by the elaboration of a disposition. We ask what the world would have to be like in order that those actions should satisfy what we know or assume to be his desires. Thus his actions are not evidence for what his beliefs are, but are constitutive of his beliefs. What a man believes depends upon what he does, and it is within our powers to determine how we act. Thus it is at least difficult and perhaps impossible for a man to act against what he believes, for to do so would be to make himself believe what he does not. In this way are constructed arguments that incontinence is impossible. And in this way Freud developed the concept of unconscious belief.

Judgments, on the other hand, may be, and are, in several well-developed theories, thought of as causes of actions. They are events which are specifiable independently of the agent's actions. On Aristotle's view a particular judgment ('This is sweet') in the presence of a general disposition to action ('Sweet things are good to eat') may lead to an action (eating this). The particular judgment may then be both an efficient cause of the action (its occurrence produces the action) and a final cause (its content may give the end of the action). In Hume's view of cause it is also possible that judgments should cause actions, since, in reasonable men of enduring character, judgments of the same sort will in similar situations usually be succeeded by actions of the same sort. On neither of these views need the connection be necessary or invariable, it resting in both cases possible and occasional that men fail to perform the consequent act, though the preconditions exist and the judgment be made. That is incontinence. Aristotle counts it as a failure of the practical reason, it is like making an inference from asserted premises and failing to assert the conclusion, the difference being that in the case of practical inference the conclusion may be an action which is not an assertion. Other writers have called it *weakness of the will*, intending thereby to indicate, as Aristotle did not, that the will has failed to act as it should. On Hume's view incontinence is just a case of the breakdown of a causal association, which is always possible.

On an Aristotelean view about actions and judgments, given a description of a general disposition to action and of a judgment, the pair forming the premises of a practical inference, we can say what the consequent action is. We have thus a clear sense to incontinence, because we have a clear sense to *consequent action*; incontinence is failing to perform the consequent action in the presence of the premises. On a Humean account, however, we have not a clear notion of consequent action, outside of an action of the sort which usually succeeds such a judgment. We have thus a trouble similar to that which arises in trying to say how a man can act against what he believes: On a Humean view to act against what one judges must be to fail to perform the action of the sort which one customarily performs succedent to such a judgment. There is no tie of content between the judgment and the action. Thus a man who has consistently failed to abstain from drinking following judgments that drinking is bad for him, will be counted as incontinent on those rare and recent occasions when he does abstain following such a judgment. The customary connection is between the judgment that drinking is bad and drinking. Thus the Humean cause of his act of drinking is that judgment. Thus the action consequent to that judgment is drinking, and he is incontinent when that action does not follow that judgment. Clearly that does not capture the sense of incontinence, and it forbids our understanding, as Aristotle's view does not, why or how incontinence is bad and should be avoided. The trouble is that the understanding of 'consequent action' as 'usually succedent action' is insufficient. Some other notion of consequent action, some way of relating actions to the propositions judged, to the contents of judgments with which they are associated, must be found. The Aristotelean view does this, and is not bothered by examples of the above sort. (It is another question whether it is possible for a man always or customarily to be incontinent with respect to actions and judgments of a given sort.)

One way of making the association is by way of beliefs. Beliefs and judgments are obviously associated by their contents, belief is a cognitive disposition and judgment a cognitive awareness both related to a content. Beliefs are associated with actions in terms of their contents, the contents being just a description of a milieu in which the actions would achieve the agent's desires, and we may thus think of the appropriate consequence of a judgment as the action which would satisfy the agent's desires in a milieu in which the content of the judgment were true. That is just to say that a man acts incontinently when he acts so as to make himself believe what he judges to be false, or to make himself fail to believe what he judges to be true.

## 0.4. Belief, Judgment and Logic

I argued above that implications and inconsistencies among propositions about judgments are few; propositions describing inconsistent judgments may be consistent. There is also apparently no uniform or plausible way of specifying what the implications are among propositions about judgments. We are fairly clear that the logical relations of propositions about judgments must depend upon the logical relations of their contents, but we are not clear how. Even in cases where we have a good account of the logical relations in a class of propositions, we may be, and in fact usually are, far from clear about the logical relations among judgments of which those propositions are the objects.

The situation is different with belief in the dispositional sense. In fact, if belief is understood completely as disposition to act, and as having no essential relation to the individual's awareness, then it is just as impossible to attribute inconsistent beliefs to a man as it would be to attribute incompatible dispositions to an inanimate object or substance. If his beliefs are just the descriptions of a world in which his actions would make sense, then any inconsistency in his beliefs becomes an inconsistency in our description of them, for we purport to be describing a possible world by giving his belief. In this extreme case, when belief is taken to be pure disposition, the logic of belief attributions is inseparable from the logic of the propositions believed, and we are unable to affirm an implication between beliefs without at the same time affirming an implication between their contents. This means that, in this sense of belief, we are quite unable to attribute a false belief to a man on the basis of his actions, for his beliefs must be tendencies or dispositions to act in the world as we see it, and we cannot consistently affirm that the world is other than we take it to be.

Thus the more we take beliefs to be given in actions, the more we take them to be public, and objective. In the limit, just described, beliefs can be inconsistent only in the sense that our description of them is inconsistent, but that is no sense at all, for an inconsistent description describes nothing.

The description of judgment, understood as a cognitive awareness, offers complementary difficulties. We are always in the situation of trying to express another's judgments in our words. We may have to avoid the words he would use were he to assert the judgment, not only if we understand no common language, but even if we attach different senses to those words than he does. Another man's judgment, if we take it to be a pure awareness, may be ineffable for us, and no description we give of it in terms of its content may be adequate. In this case his judgment would be completely opaque to us, and we could affirm no logical relations among various judgments which might be his. Logical relations depend upon structure, and we have no way of discovering the structure of another's awareness.

### 0.5. A RELATIVE VIEW OF BELIEF AND JUDGMENT

Neither of these extremes, the view of belief as a pure parameter of action, nor that of judgment as a moment of awareness, is adequate. In the first case we have a notion of belief which ignores the awareness of the agent, and in the second a notion of judgment to which consistency does not apply. But there are concepts between these extremes which are useful. One way of distinguishing these concepts among themselves is by the extent to which the logic of the contents determines the logic of the judgments or beliefs. When there is no determination whatever, we have the notion of judgment as pure awareness. When we allow different sorts and extents of determination, we obtain concepts which are distinguished by the ways and extents to which they are public and objective. When the determination is total, we have the completely public concept of a parameter of action.

In terms of this scheme, we may, in considering the cognitive attitudes of a given agent, take his judgments and his beliefs to be related in this way: The logic of his beliefs is at least as much determined by the logic of their contents as is the logic of his judgments determined by the logic of their contents. This allows us to associate the judgments and beliefs of the same agent, and also to understand incontinence: The man who is incontinent makes a judgment and fails to act in accordance with it. That is to say, he judges a proposition which he does not believe. If we did not distinguish his judgments from his beliefs, they would be inconsistent, and, assuming a modicum of awareness on his part as to what he believes, if we allowed inferences among his judgments by all the logic used in relating his beliefs, his judgments would be inconsistent. But we distinguish his judgments from his beliefs by attributing a weaker logic to the former, and we have thus consistent descriptions of his awareness (to the extent that we take it to be describable) and of the public features of his cognitive life as it is revealed in his action. It is perhaps easier to say that belief and judgment give us distinct views of the reality which is his cognitive attitudes, variously expressed.

In this way a man may act incontinently without making inconsistent judgments or having inconsistent beliefs. His failing will, however, be a sort of a logical error, namely a failure to apply a sufficiently strong logic to his judgments. That is much like what Aristotle said about it, with a slight difference. He described the incontinent man as failing to apply a logical rule, and the present view does so too. The difference is that on the present view we do not insist on the uniqueness of a system of logical rules, and it is allowed that one man may have, or commonly employ, logical rules which another man does not. Aristotle took everyone to have the same logic and he was thus, in the case of incontinence, apparently puzzled by the problem of how a man could be said to have a rule which he failed to employ upon an obvious and proper occasion. The present account does not resolve this puzzlement, but attempts to avoid it by relativizing the notion of a logical rule.

It is an advantage of this view that according to it it is theoretically possible for a man always to be incontinent. Whether it be psychologically or morally possible, that is to say, whether we should deny that such a man is an agent on the ground that he is mad, or bereft of moral sense, is another matter. But it ought not be a consequence of an account of incontinence, as it is of the Humean account which I sketched above (that account is not Hume's, but is done in his terms) that continued or customary incontinence is a contradiction in terms.

There are several things to remark about the theory put forth here. First, it is described only very roughly and dogmatically and is not developed in any of its details. That is partly because I have thought about only a few of those details, for the most part those which have to do with its application to partial beliefs, and I am not as confident of its adequacy as I should hope to be upon further development. But I wanted also to make some remarks in brief compass to give some idea of the general setting in which the views of the succeeding chapters might be understood.

Very little is said about desire here, and the question of relation of desire to action is put aside in the reference to dispositions or principles. This is inadequate, but may be excusable in view of the attempt to address as directly as possible the question of the nature of belief. I should not presume to develop a general theory of action, belief and desire, and I have mainly been concerned that the development with respect to belief should contradict currently received views about action and desire as little as possible, and then just in respects in which they seem wrong.

Something should be said as well about the relations to the views of Quine and Davidson. The problem of understanding another's judgments is quite like the problem of understanding another's assertions, which Quine has called the problem of radical translation. He has showed how intimately this is related to the question of the nature of analytic necessity. In particular he has made it clear how our views about understanding assertions involve essential and viciously obscure assumptions about necessity. The response to this which Quine seems to favor is the attempt to develop accounts of comprehension without reference to the concept of necessity. I have taken the easier option of retaining reference to necessity, which reference is kept as far as possible explicit, so that the account of belief is obviously relativized. It is important to mark this difference for the following reason: Quine's arguments do not undercut all use of a concept of necessity in giving an account of comprehension. Insofar as we may give a plausible clear meaning to necessity there is no essential difficulty introduced in an account of comprehension by dependence upon necessity. Thus, for example, first order logical truth is sufficiently clear to be employed in this way. Further, the difficulties introduced by less clear concepts are not merely those consequent upon the adoption of certain minimal standards of clarity which the concepts in question fail to meet. Were that the case, the Quinean arguments would have shown no more than that certain, even all presently developed, accounts of comprehension, conflict with the adoption of those standards, and this could as well be taken as a difficulty with the standards as with the accounts. More than this is at stake, however, for the arguments are directed at showing that the unclarity introduced by unanalyzed concepts of necessity is in general vicious, in the sense that it assumes the resolution of the very issues in question. This point is important, and, though frequently insisted upon by Quine, is sometimes ignored by his critics.

The decision to retain reference to necessity without restriction as to the nature of the concept in the discussion of comprehension was not made in disdain of Quine's arguments, but was forced by the apparent impossibility of describing the relation of probability and belief without employing notions of necessity. This would cause no trouble were it not fairly clear that those notions are not antecedently understood, and may, in quite standard cases, introduce just the sort of impredicativity which Quine has pointed out. This does not seem to me to be an accident of the approach that I have taken, nor do Quine's arguments seem to lose their point.

The above discussion of incontinence owes much to Davidson's paper on the subject. He addressed the problem of incontinence directly in that paper, while I came to it from the question of the relation of belief and action. Davidson accounts for the possibility of incontinence by allowing that the agent may hold principles which he does not, on a given occasion, employ, even though hc recognizes their applicability to the case at hand. I have distinguished principles employed from principles merely held in terms of the distinction between belief and judgment. This done, Davidson's resolution seems right to me, and I have tried to ornament it by asking as well how beliefs and judgments are related. This has turned into the question of how beliefs understood in terms of conduct are related to the agent's cognitive awareness, and that question has been treated in terms of differences between their logics.

Davidson should also be credited with insisting on the difficulty of giving a consistent description of inconsistent beliefs, and on that of separating the evidence against a proposition from the evidence that a man does not believe it. He has not done that explicitly in print, but the influence is pervasive throughout this book as well as in more important places.

CHAPTER I

# THE NATURES OF JUDGMENT AND BELIEF

## I.1. Remarks on the Theory of Judgment

It is natural and usual in a theory of judgment to distinguish the *content* of a judgment, what is judged, usually a proposition, from the act of judging. There are theories, most notably Hume's, in which this is not done. The difficulties which such theories encounter result for the most part from the difficulty they have in allowing the mind to entertain or assume the same proposition which it might also judge. I postpone until succeeding sections the question of how such theories may be modified to meet these problems and turn initially to theories in which the act is distinguished from the content.

Given that we make the distinction of act from content, the question will arise whenever we consider a feature of judgment, whether we should account for it as a feature of the content, or a feature of the act, or as some combination of these. Consider, for example, negations. We may think of judging *A* and judging not-*A* in either of two ways:

(1) They consist of the same act with different contents.
(2) They have the same content which is acted upon differently in the two cases.

If we take the first alternative, then we should presumably take the logical relations among judgments to be the logical relations among their contents, and on this ground account for the inconsistency of these two judgments as resulting from the inconsistency of their contents. By a natural extension, defining equivalence in a usual way, we have, for example, that the judgment that *A* is equivalent to the judgment that not – not-*A*, and we have the question how these two judgments can be distinguished or whether indeed they are the same judgment differently described, as would be, for example, the judgment that there are non-denumerable infinities and the conclusion of Cantor's famous diagonal argument. In this latter case the only distinction to be made is in our descriptions of the judgment. And if the logical relations among judgments are to be just the logical relations among their contents we run the danger of not being able to distinguish double negations.

This particular example may not be too distressing; if one has a sufficiently classical attitude toward negation then he takes even numbers of negations to be simple affirmation and any odd number of negations to be simple negation. But it is not hard to extend the argument by means of other logical relations to cases where this option is totally implausible, and the ultimate consequence is the complete transparency of judgment, leaving us incapable, for example, of drawing any distinction among judgments in mathematically distinct propositions.

This is not an argument that any way of relating judgments in terms of the logical relations of their contents must have this consequence. I am just pointing out a difficulty which such ways must take some care to avoid. The true theory of judgment must in at least some cases be able to distinguish judgments with equivalent contents.

Such considerations may lead to making the distinction between judging *A* and judging not-*A* in the second of the ways mentioned above, as consisting of different acts exercised upon the same content. In particular, the second is said to be a denial of *A*.

Making the distinction in this way, though it distinguishes the two judgments, says nothing at all about how they are distinguished or related to each other and to other acts of the mind such as assuming or questioning. Nor is this simply a problem of filling in an account. It is quite clear that judging *A* and denying *A* are incompatible in ways, for example, that assuming *A* and desiring *A* are not, and it will not be easy to say in what this incompatibility consists and at the same time avoid the very total transparency which the account was designed to avoid.

A third alternative is worth mentioning though it seems pretty clearly unfruitful. This is to suppose that judging *A* and judging not-*A* consist of the same act with different contents, and that the contents are not related in any uniform way; the distinction to be made is just that between judging *A* and judging any *B*, where *B* is a distinct proposition. It is sometimes said that only increased research in neurology or behavioristic psychology will be able to uncover the relations between such judgments. This alternative seems unpromising because it ties judgment too closely either to language or to behavior. If the former, then the way in which the judgment is identified will depend upon the language in which it is reported and the account will be at bottom not neurological at all, but linguistic; in which case the relations among judgments will be determined by the relations among the sentences which express them, and the question whether two judgments are compatible will reduce to the question whether the assertions which express them are compatible, which is apparently just another form of the same problem. It is sometimes said that any two assertions are compatible, in the sense that any string of vocables may be uttered by one with the proper physiology and inclination. This seems to me to confuse utterance and assertion and to amount to no more than a refusal to see the problem.

The other option, of tying judgment to behavior, has the difficulty in an extreme form that judgment is completely transparent; *we* describe the behavior, not the judger, and of course distinct descriptions may be true of the same behavior. The alternative to this is to eschew extensional descriptions of behavior, and to insist that the proper description fit the behavior as intentional, or from the agent's point of view, which of course leads one directly to the problem of the nature of judgment and cannot be counted as a solution at all.

In any event, linking judgment to behavior means that incontinent behavior, in which one acts contrary to his judgment or desire, becomes impossible, supposing of course that desire is also interpreted behavioristically. If what it means to make a judgment is to act as if the content of the judgment were true, then a man could never fail to act on his judgment, which is to say, he could never fail to act as if his judgment were true. This holds whether we think of behavior in narrow extensional terms or as including reference to the agent's view.

To summarize: Neither the distinction of judgments in terms of acts nor in terms of contents will be easy. Both will have troubles in allowing some transparency to judgment; enough, say, to count judging *A* and judging not-*A* as incompatible, without allowing too much transparency to judgment, with such results as that judgments in mathematically equivalent propositions cannot be distinguished.

There is sometimes a tendency, particularly when the difficulty of the subject is brought home to us, to try to treat the problems of the theory of judgment in the philosophy of language rather than in their own right.[1] I see nothing wrong with this tendency, for precisely the same questions will come up there. As soon as one turns to what is now called depth grammar, or depth semantics, all of the questions we have been considering arise as questions, for instance, about the semantical role of negation in English.

Another point of view disparages not only the theory of judgment, but semantics as well, and turns its attention to accurate descriptions of the uses to which sentences are or may be put. This approach seems to me to lead shortly back to the theory of judgment for the reason, roughly put, that it is not in general possible to describe the uses of sentences in any rewarding way without becoming involved in questions of the nature of their truth conditions. I cannot insist upon this, for it is not a point which I can presently argue, though it seems that there should be a good argument for it. It is rather the sort of thing of which one becomes convinced in the application. That is to say, the more that one tries to give adequate accounts of how sentences are used, the more one sees that the question of truth conditions and in some cases of reference is quite central to this.

There is one more topic in the general theory of judgment which it may be helpful to discuss briefly now, somewhat in anticipation of more detailed discussions later, before turning as I shall to modalities and the *de dicto – de re* distinction. This is the topic of partial judgment. Assuming, as we have been, the distinction of act from content the phenomena of partial belief are subject to the same question as that raised about negations earlier. That is to say, are we to think of partial judgment as differing from apodictic or non-partial judgment in its act or in its content or both? In the first case, we shall say that partial judgment amounts to taking a judging attitude of less than full strength toward a proposition or content which one could on another occasion judge fully, apodictally, or non-partially. In the second case we should strive to account for partial judgment in terms of some partiality not in the act but in the content, the act being supposed to remain the same as in straightforward non-partial judgment. Interesting and worthy theories have been developed in both these directions. Hume allowed for both sorts of partial judgment.[2] Though he pretended not to distinguish the act from the content, it is very difficult to make out his accounts of probability and of chance without distinguishing full belief in frequency propositions, in which the probability is made part of the content, from partial belief, in which the mind in accordance with the principle of indifference spreads its conviction over alternatives which in themselves need not involve probability at all. The second of these theories, as dependent upon vague metaphor as it is, still seems to me not to have been improved upon as an account of partial acts of judgment. The first, according to which partial judgment involves always partiality of the content and not of the act, if held in the absence of the first theory encounters great difficulty (1) in the case of judgments where no frequency propositions can be found, as in the case of a man who bet on whether the first manned moon rocket would arrive safely at its destination, and (2) in cases where there are several incompatible frequency propositions, as in the case of a man who judges about the chance of an accident in his automobile. He will judge differently depending upon whether we take the appropriate reference class to be, for example, trips of a given length in miles, or in time, and whether we count as relevant the type of car, or the condition of the traffic, or what have you. Thus when we say that he judges to a certain extent that he will have an accident on a given trip, we cannot in general give the frequency proposition which he believes fully, unless we allow that it is whichever proposition may be obtained by abstracting from the proposition in question which has the suitable frequency value, that is to say, unless we allow the frequentist first to know the value of the frequency he is looking for, which is pretty clearly no theory at all.

One of the important things to see is the general relevance of questions about the distinction of act from content in judgment. It is almost always of considerable importance whether we count a given feature as a feature of content or as a feature of act or as having the capacity to arise in either place. In the latter instance we face the question of the relation between the judgments of the two sorts; between denying *A* and affirming not-*A*, or between a full belief in a frequency and a partial belief in a simple non-frequency proposition.

What I have been calling the different acts of judgment many philosophers have thought of as an act which is susceptible of different modes. This is in no way incompatible with the sketch I have given; I chose to make the distinction between act and content for expository reasons. Speaking about *modalities* leaves open the question whether they are to be viewed as act or as content, as the development of modal logics makes clear. In these logics the modalities are considered always as being capable of occurring in the content. The reason for this is worth remarking since it emphasizes another question of interest in the theory of judgment: Accounting for a modality, say, necessity, as being involved only in the act and never in the content leaves the explanation of embedded occurrences of necessity, as in the judgment that a proposition is necessarily entailed by its necessitation, quite unexplained. Necessity in the act may make sense of judgments of the form '*A* is necessary' but once the necessity becomes embedded, particularly within the scope of operators such as the truth functions, which may not be given an interpretation as part of the act, but rather included in the content, the intuition, at the bottom of the analysis into act and content, that these must be in every case clearly separable, is no longer clear enough to sustain this account. At least since the publication of *Principia Mathematica* the objects of judgment have been taken to have a recursive structure, the procedures for composing new contents out of old permit reiterated applications of the results of former applications. And, indeed, it is largely through this feature that the infinity of possible contents and their relations may be brought under theoretical control. The act, on the other hand, when we distinguish it from the content, has no such apparent structure, and we do not have much intuition at all that there should be a recursive algebra of acts. For these reasons modal logics, which depend upon a recursively structured semantics, analyze necessity as a part of content.

The embedding problem arises also for partial judgment, and an account such as Hume's according to which partial judgment consists of spreading out the force of the mind over distinct contents, must come up short upon the occurrence of probability as an embedded constituent of judgment. Indeed, the science of statistics abounds with examples of partial judgments about probabilities. The latter, as examples easily show, cannot easily be thought of as frequencies, nor can they be analyzed in any apparent way as some feature of the act. This situation is, I think, one of the most difficult in the philosophical theory of probability, and I have no good idea in which direction a solution is to be sought.

Let me turn now to the distinction of *de dicto* from *de re* propositions. The theory of judgment which was most prevalent from the time of Aristotle until the last few decades[3] was that simple attributions of a character to an individual were always *de re*, which was then taken to mean, in our terms, that the individual substance itself occurred in the content and was not merely conceived of in the judgment. The judgment that Caesar died is just a predication or attribution of death to Caesar. This is the view which Russell held in the *Principles of Mathematics* and which, at least as far as judgments expressed by sentences including proper names as subjects are concerned, he never completely abandoned. When the theory, at least up to the time of Frege, turned to the analysis of judgments not of this simple form, such as general propositions expressed by sentences with quantified expressions as grammatical subjects, it encountered the difficulties which led to the theory of suppositions and, in Russell's early work, to the account of denoting given in the *Principles of Mathematics*, from which the notion of incomplete symbol and the concomitant account of definite descriptions is absent. The problem was to find a referent for quantified expressions, such as 'all men' which grammatically function on the surface quite like proper names. Or, to put this another way, the problem was to find what constituent of the proposition *All men are mortal* is all men. Russell treated this problem, in the *Principles*, by supposing that there was no such constituent of the proposition, but that instead the *concept* of all men occurred in this proposition and that this concept stood in the relation he called denoting to all men, which he took to consist of the members of the class of men conjoined in a special way. The need to introduce a notion such as denoting is even clearer in the case of propositions such as *Some men are Greeks and some are not*. We cannot suppose that some men are constituents of the proposition, for they cannot be both Greek and not Greek. The notion of denoting was introduced to allow the concept to stand for an entity (the subclasses of the class of men, disjunctively considered) which would make proper sense of the proposition.

In view of this we can see how general propositions could not, before the analysis of such propositions in terms of quantification theory as being essentially pronominal, be viewed as *de re* at all. No individual or collection of them was included in such propositions. Thus, as Kaplan[4] emphasizes in his recent work on proper names in which he returns to a theory much like that of Russell in the *Principles*, the distinction between *All men are necessarily mortal* and *It is necessary that all men are mortal* cannot properly be said to be one between a *de re* and a *de dicto* proposition at all. It is much more clearly understood as a difference of the scope of the modality. In the one case this scope is a propositional function and in the other it is a proposition. On the other hand, sentences with a proper name as subject could never fail to express a *de re* proposition, whether that proposition involved a modality or not. Both *Caesar is necessarily mortal* and *It is necessary that Caesar is mortal* are *de re*. Caesar is a constituent of both of them. They differ, as do the examples above, not with respect to the sort of modality but with respect to its scope.

One important thing to notice about the *de re*, *de dicto* distinction is that it is not to be made in terms of scope. Indeed, a proposition such as *Caesar died* is *de re* though no question of scope arises in its analysis.

Another thing to notice is that the distinction is not properly made in terms of referential transparency. In order to examine this question we must look directly not at the propositions but at the sentences which express them.

One distinction between

(i) Scott is the author of Waverley.
(ii) George IV believed Scott is the author of Waverley.

is that the former but not the latter is referentially transparent at the position marked by 'the author of Waverley'. On the Russell–Kaplan view of proper names, both (i) and (ii) express *de re* propositions, the first with respect to Scott, and the second with respect to Scott and George IV (assuming 'George IV' to be a proper name and not a description). It is clear that sentences which express *de re* propositions are transparent at their subject positions as far as the preservation of truth value is concerned. This invites us to ask if we can make all the distinctions we need without bringing in propositions at all, and strictly in terms of the referential transparency of specified positions in sentences, all of which are construed as *de dicto*. To express a *de re* proposition, one would use a *de dicto* sentence and mark within it the position intended to be purely referential.

First, as far as doing away with the use of propositions in general, I do not think that that can be accomplished as far as the theory of judgment is concerned, because I see no way in which an analysis of judgment can be accomplished without distinguishing act from content. Independently of this, however, and assuming the correctness of the Russell–Kaplan view of proper names, it is still an interesting question how the distinction is to be made among sentences, and one way of trying to do this is in terms of referential transparency.

An argument due originally to Frege and since used in various connections by Church, Quine and Davidson,[5] will show, if it can be properly made out, that this will not be easy, for the reason that it is not easy to restrict referential transparency to a position within a sentence. Let me first say how the argument is supposed to go and then make the premises explicit.

Suppose we consider a modality $M$ and a sentence

(i) $M.F(a)$

which we intend to be referentially transparent at the subject position. Then, assuming an ordinary logic and theory of descriptions, (i) is equivalent to

(ii) $M.F(Ix.x = a \wedge F(a))$.

Now let $G$ be any open sentence which is coextensive with $F$. Then since

$$F(a) \leftrightarrow G(a)$$

(ii) is equivalent to

(iii) $M.F(Ix.x = a \wedge G(a))$

and since $F(Ix.x{=}a \wedge G(a))$ and $G(Ix.x{=}a \wedge F(a))$ are logically equivalent, so long as we assume that the modality $M$ does not distinguish among logically equivalent sentences, we have that (ii) is equivalent to

(iv) $M.G(Ix.x = a \wedge F(a))$

which is equivalent, by transparency of the subject position, to

(v) $M.G(Ix.x = a \wedge G(a))$

and thus to

(vi) $M.G(a)$.

This argument shows that transparency for coextensive replacement cannot be restricted to a given position in a sentence; if subject positions are transparent in this way then so long as the modality in question is transparent for logical equivalents it must also be transparent for replacement of coextensive predicates. It will help in understanding this argument to make some of its assumptions explicit. In particular we need

(I) If $A$ and $B$ are logically equivalent then $M.A$ and $M.B$ have the same truth value;
(II) $F(Ix.x = a \wedge F(a))$ is logically equivalent to $F(a)$;
(III) $F(Ix.x = a \wedge G(a))$ is logically equivalent to $G(Ix.x = a \wedge F(a))$;
(IV) If $F$ and $G$ are coextensive then $(Ix.x = a \wedge G(a)) = (Ix.x = a \wedge F(a))$

and these assumptions permit argument from: (1) $M.F(a)$ is transparent at the subject position; to (2) $M.F(a)$ is transparent for coextensive predicates at the predicate position.

The conclusion I think we should draw from this is that the distinction between *de re* and *de dicto* sentences is not properly made in terms of the referential transparency of subject position. This conclusion is reinforced when we notice how little of assumption (I) is used in the argument; the modality needs to be invariant only for quite weak logical equivalence in order for the argument to work. Thus we seem to have needs for distinguishing: (i) *de re* from *de dicto* propositions, and both of these from (ii) referential transparency of modalities, and (iii) scopes of modalities. And it does not seem that any of these will collapse into the others. As a consequence of this, we need to recognize at least two forms of a given modality, one which applies to propositions and the other to propositional functions.

## I.2. MENTALISTIC VIEWS OF BELIEF

We turn now briefly to questions of the *psychology* of belief; to consideration of believing. There are classically two accounts of this, not necessarily incompatible. One of these is *mentalistic*, according to which belief is a mental occurrence in the believer. The second is *behavioristic*, according to which a belief is a parameter of action, or, as Russell puts it, like a force in physics, is a fictive cause of a sequence of actions. We consider these accounts in turn.

The most famous mentalistic account of belief is that of Hume according to which belief is just lively conception. This account has several important features:

(1) The form of belief. The most natural form of belief on Hume's view is belief of or in an object. To believe that an object has a property is either to have a lively conception of that object, which conception includes a lively conception of that property as a part, or else to have a lively conception of that object followed by a lively conception of the property. Thus the ordinary predicative form '$a$ is $\varphi$' is reduced either to essential predication, when $a$ is not conceived without $\varphi$, or to causal predication, when the concept of $\varphi$ regularly follows that of $a$.

Hume gives no explicit account of compound beliefs; of beliefs in logically complex propositions, for example. In some cases it is fairly clear how he would proceed.[6] He would almost certainly count belief in $A$ and $B$ as just belief in $A$ and in $B$, thus taking belief to have a sort of transparency for conjunction. This is plausible, but other truth-functional compounds work not nearly so well, disjunction being an obvious case. Certainly, believing $A$ or $B$ is not just believing $A$ or believing $B$: It is not sufficient for one might believe that rain will fall or the sun will shine without believing either. (If one is an intuitionist then belief in $A$ or $B$ entails belief in $A$ or belief in $B$. The intuitionists, however, have been careful always to restrict such doctrines to mathematical propositions, where counterinstances are not so easy to come by.) Nor is it necessary, for believing $A$ – having a lively conception of $A$ – does not entail for every $B$ having a lively conception of $A$ or $B$. If it did then either (i) every belief would entail an infinite number of distinct disjunctive beliefs, or (ii) disjunctive beliefs with a common disjunct would not be distinct.

Similar difficulties arise with negation; believing not-*A* cannot be just not believing *A*.

One way of dealing with these problems is to treat logical concepts as substantives, taking, for example, belief in *A* or *B* to be a lively conception of; *A*, or, and *B*; and taking belief in not-*A* to be lively conception of not and of *A*. Thus conjunction becomes the only mode of combining beliefs, and belief is transparent for it in the manner described above. The other truth-functions become intentional objects in their own rights. This view, which is very implausible at first, becomes much less so after reflection.

There is this difficulty. Believing *A* or not-*B* cannot be simply believing (having a lively conception of) *A*, or, not, *B*, for we could not then distinguish it from believing not-*A* or *B*, or believing not (*A* or *B*). Thus if we are to include the logical operations among the objects of belief, we must include also some indexing or ordering of them. We should, for example, have to take believing *A* or not-*B*, to be believing *A*, believing or, believing not, and believing *B*, in that order. Even this, however, does not permit us to distinguish believing not-*A* or *B* from believing not (*A* or *B*). To make this distinction we might suppose something like the Polish notation[7] to be innate, so that believing not-*A* or *B*, would be believing or, believing not, believing *A*, and believing *B*, while believing not (*A* or *B*) would be believing not, believing or, believing *A* and believing *B*. Alternatively the object of belief might be not a sequence but a structure;

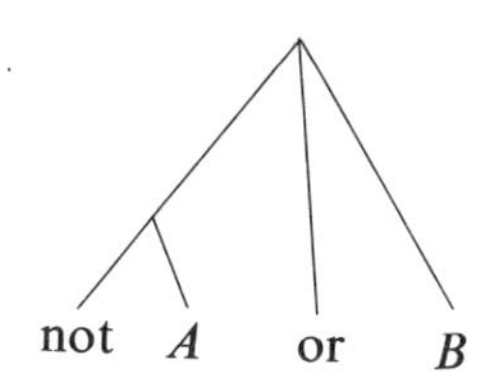


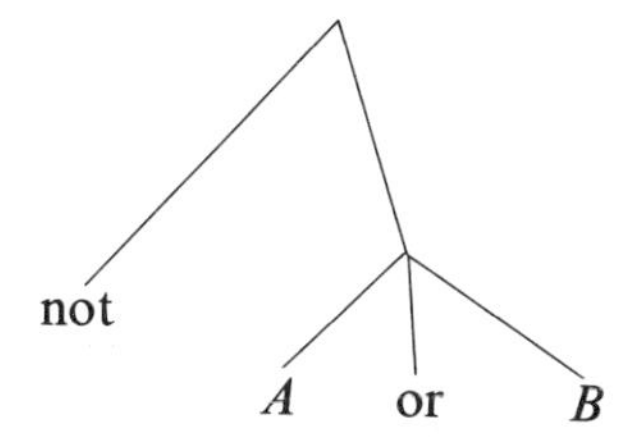


but in either case we have not reduced belief in a compound to belief in its components, even when the logical ideas are included among these components, for in each case we require the superposition of some structure on these components.

The case gets even more difficult with quantificational notions. It is not very plausible to suppose that believing $\exists x A(x)$ is believing of some object

$a$, $A(a)$ : I may, in the case of scientific laws, for example, believe $\forall x(A(x) \rightarrow B(x))$, intending this to apply to objects $a$ which I have never considered.[8]

One way of trying to meet this is to treat laws as expressing a relation among properties, that is to say, as being of the form 'The property *A* implies the property *B*'. And to treat beliefs in other universally quantified compounds as limits of beliefs involving their instances. On mentalistic accounts this treatment of laws seems promising, but the reduction of beliefs in quantified compounds to beliefs in their instances seems much less so for the reason that it requires unreasonable assumptions about the number of intentional objects before the mind of the believer.

In general, mentalistic views may try to account for beliefs in compounds either through devices of transparency, reducing, for example, belief in *A* and *B* to belief in *A* and belief in *B*; or through the substantivization of the connectives which form the compounds from their components. Hume had a strong tendency to move in the former way; his account of spatial and temporal beliefs is of this sort.[9] He thinks of believing *A* precedes *B* as believing *A* preceding believing *B*; and, it is pretty clear, of believing that *A* is to the left of *B* as involving the spatial relation of being to the left of between the conception of *A* and that of *B*. He also thought of himself as analyzing belief in *A* causes *B* into the belief in *A* causing the belief in *B*.[10] These attempts have their fairly obvious difficulties: As far as temporal precedence is concerned it becomes impossible for me to be uncertain about temporal precedence; if I have a belief in *A* and then a belief in *B*, it follows that I believe that *A* precedes *B*. Also, for any *A* and *B* which I believe nonsimultaneously, I must either believe that *A* precedes *B* or believe that *B* precedes *A*.

(2) Another problem for Hume's form of mentalism concerns beliefs of the form '*a* does not exist', which, on his view, are impossible. This is because, for Hume, to conceive of anything and to conceive of it as existing are the same thing. Hume is led to this because he can find no impression or conception of existence distinct from the impression of a particular object.

This consequence of Hume's view is pretty clearly undesirable. The question arises to what extent it is essential to the mentalistic view of belief.

The problem is to give an adequate mentalistic account of beliefs of nonexistence. Given that existence is a property of objects, Hume's claim that it cannot be separated in thought from the idea of an object is quite plausible, though of course not incontestable. Thus if we are to provide a non-trivial account of judgments of nonexistence in line with Hume's views we should look at the possibility of analyzing existence in some other way. Such an account is given by Russell,[11] according to whom singular existence is a *second-order* property, a property of properties, and is predicable of individuals only derivatively and intentionally. A property is said, in Russell's view, to have singular existence if exactly one thing has the property. To say that the property $\varphi$ has singular existence is to say that there is some object such that it and only it has $\varphi$. Thus the notion of singular existence of properties is analyzed into components of identity, generality, and truth functional combination. In view of this an object $a$ can be said to exist only derivatively and with respect to some property of which it is the unique exemplar. $a$ exists *qua* $\varphi$ if $a$ is the unique $\varphi$. This means that judgments of nonexistence are general judgments, to judge that the $\varphi$ does not exist is to judge that either there are no $\varphi$'s or more than one.

This does take care of the difficulties of Hume's view, but at the expense of relying on judgments which are essentially general in nature. Hume's project was, insofar as possible, to take judgment to have always a particular object or a finite number of particular objects, and he did not envision properties as being among these objects. But that part of Hume's view can, at least initially, be separated from the mentalistic account of belief, and as far as modifications in that theory are concerned, Russell's account is legitimate and effective. The Russell account does emphasize the importance to a mentalistic theory of accounting for quantificational compounds as objects of belief, for existential propositions now become compounds of this sort, but this is no disadvantage of the view, for these compounds must be accounted for anyway.

(3) Another obstacle faced by Hume's theory, and indeed by any mentalistic theory of belief, is that of accounting for the non-episodic sense of belief, according to which a man has beliefs of which he may at any time be unaware. A large number of these beliefs may, on Hume's account, be thought of as habits of mental action. Causal beliefs are of this sort; to say that I believe that $\varphi$'s cause $\psi$'s is to say that were I to have a belief that something is a $\varphi$, I should then believe also that it is a $\psi$. Such beliefs are thus *a priori* in a very real sense; they never figure as explicit objects of my consideration, but are themselves regularities which govern my mental life. This way of dealing with dispositional belief is quite general in Hume, he tries as far as possible to account for a belief of which the subject is at the moment unaware as a tendency of which he

may never be aware. What is most surprising about this is the unexpectedly large number of such beliefs which are adequately dealt with by it.

Beliefs of the dispositional sort which are not accounted for by tendencies to believe may sometimes be accounted for as analytic. Most mathematical beliefs will be considered in this way. Here one encounters problems similar to those involved in questions about logical compounds; that is, it is difficult to maintain the mentalistic feature of the theory. We may say, for example, by fiat, that a man believes all the mathematical consequences of his beliefs. Thus guaranteeing that a man who believes there to be nine planets also believes there to be more than eight planets, but this means, if we persist in this account, that Hobbes believed that the circle could not be squared. Of course, *this* consequence may not be undesirable, since Hobbes had a mathematically inconsistent belief, but it also follows that Hobbes, or anyone else who has an inconsistent belief, believes everything, in spite of his convictions to the contrary, and this consequence seems to deprive the mentalistic account of the plausibility which initially recommends it. We must, if we are to save that view of belief, be able to separate beliefs in quite closely related propositions.

One way to do this is by using another sort of dispositionality: To say, of a mathematical proposition, for example, that an individual believes it (in the dispositional sense) is to say that, were he to consider it he would believe it in the non-dispositional sense. This is, however, impossible on Hume's view since belief is in his theory just lively conception, and thus no line can be drawn between consideration and belief.[12] For this reason, *reductio* proof is, on Hume's view, difficult to account for. In such proof, one assumes some proposition in order to derive its denial, thus establishing that the assumption is false. But on Hume's view, every such proof would require *believing* the assumed premise, since it would be clearly conceived. This is plainly not the case. Further, on this view, it is not clear how such proof differs from mere change in beliefs; one starts off believing *A*, ends up believing not-*A*; and is not in any proper sense proof at all.

The resolution of this suggested by Hume's account of *demonstration* in the *Treatise*[13] is to treat the proof as a whole as one judgment, of the form, in the case of *reductio*, 'If *A* then not-*A*'. This has to recommend it that it makes it clear how in *reductio* the denial of the assumption is established, but it has the disadvantage that, in *reductio* in particular, a man will perforce judge inconsistently: If belief is lively conception, then he must believe the assumed premise, since he conceives of it clearly. For the same reasons he must believe the denial of that premise. This difficulty is indeed a general one, and is applicable to negative judgments in general: To judge the denial of *A* is to conceive clearly of *A* and thus to judge *A*. To resolve this problem, which is clearly a serious one, there are the following apparent strategies. (i) To allow different acts of the mind, one of which is belief or judgment, so that a proposition may be clearly conceived without being believed. This view is considered in the next section. (ii) To put restrictions on what may be an object of the mind, restricting, for example, consideration to one object at a time. Thus, in the case of a denial, the proposition denied would not be said to be clearly conceived, only the complete denial, of which the proposition is a part, would be clearly conceived, and thus only the denial would be judged. We might put this in terms of a restriction on objects: the mind may have at most one object at a time. Should it have an object, distinctly conceived, that object is judged or believed.

This is not completely implausible. In the case of *reductio* proof, for example, we should say that in considering the premise no judgment is being made, or judgment is being suspended, and that no complete object is before the mind. In addition, however, to having a somewhat *ad hoc* character, the notion of *object* of the mind now assuming an important technical status, not yet explained, it has also the disadvantage that perhaps the most plausible of the inferential rules governing belief, which is initially supported by the mentalistic view, now must be abandoned. This is the rule which allows inference from belief in a conjunction to belief in each conjunct. If we allow at most one object at a time to be before the mind, and restrict the sufficiency of lively conception for belief to objects, components of the object not being believed, then the conjuncts, no matter how forcefully conceived, cannot be believed at the same time as the conjunction of which they are members

Indeed, the alternative of allowing only one object to be before the mind at a time and only objects to be judged, amounts to a special case of allowing there to be different acts of the mind, and requires us to abandon the Humean assumption that there is only one act of the mind.

## I.3. Mentalistic views; distinct acts of the mind are possible

Another form of mentalistic account is that belief is a mental attitude toward a proposition, but that other attitudes are possible toward the same objects. This view was held by Meinong[14] after he had studied Hume with some care. It has a number of advantages over Hume's view, not the least of which is initial intuitive appeal, since we do feel that, for example, judging and assuming are distinct mental acts which may have the same content.

Before looking at this view it may be useful to see why Hume opposed it, though he must have been aware of its advantages.

The most obvious reason behind his opposition is his strong conviction that the mind cannot be distinguished from the collection of its contents. If there is no more to the mind than its propositions,[15] then clearly there can be no act of the mind save the ultimate one of including the object as content. It is clear that if we move to a mentalistic theory according to which there are distinct acts of the mind toward or involving the same contents, then we must suppose that there is more to the mind than those contents. This is almost certainly what led Hume to deny the diversity of acts of the mind, for he was emphatic in his denial of the existence of an independent mind. It is probable that he associated problems about the nature of the mind with problems about self-identity and the nature of the self of the reason and imagination. It is not, however, apparent to what extent this association is justified and to what extent, for example, one could argue from the existence of distinct acts of the mind to the existence of the transcendental ego.

Russell, in considering belief, was careful to distinguish the *act* of believing from the *content*, what is believed, and these two from the *object*,[16] the states of affairs which determine the truth-value of the belief. For Hume the content was always an impression or idea, and the act was either the occurrence of this content or the manner of the occurrence. Hume refused to speculate about the object, except in the case of impressions and ideas of reflexion. The question arises to what extent the difficulties encountered by Hume's view because of his refusal to allow different sorts of acts are met by a theory, such as Meinong's, which supposes there to be distinct acts with the same content. In particular, Meinong's theory allows that the same content may be either judged or *assumed*,[17] which latter act, roughly put, is much like judgment but lacks conviction of the truth of the content.

It would be hard to make the distinction between judgment and assumption without reference to the object or truth-conditions of the belief, and this is perhaps another reason why the distinction is made obscurely, if at all, by Hume.

Now the characteristic difficulties in accounting for *reductio* proofs and negative judgments, which seem inevitable for a mentalistic theory which does not distinguish assumption from judgment, will not arise. In a *reductio* proof we may *assume*, without conviction, a premise, and argue to a conclusion which we know to be false. This argument may lead us to *judge* the denial of the premise. Similarly, in the case of believing a negation, we may allow what is negated, a part of the content, to be before the mind, and think of the judgment, which carries conviction with it, as applying only to the whole content judged.

Mentalistic theories which allow distinct sorts of acts of the mind allow also for a different sort of treatment of logical connectives than do theories such as Hume's:[18] We may think of judging not-*A*, for example, as judging *A* negatively, or, simply, as denying *A*, where denying and affirming are different species or judgment. Thus negation is viewed as a feature of the act of judging rather than as a feature of the content judged. Of course this treatment can be extended also to the other logical particles; to judge *A* and *B* would be to judge the complex content *A*,*B* conjunctively, etc. Russell thought at one time that *tense* should also be handled in this way and that differences in tense in judgment were at bottom differences in the way of judging the same content,[19] typically differences among memory, perception and anticipation.[20]

These views encounter the serious difficulty that they do not allow for imbedding of logical particles: Judging not-*A* may be denying *A*, and judging *A* and *B* may be conjoining *A*,*B*, but this does not account for judging *A* and not-*B*. There are apparently two alternatives here. One is to multiply acts of the mind, so that in addition to affirming, denying and conjoining we have also a new form of judging, and judging *A*,*B* in that way amounts to affirming *A* and not-*B*. This alternative is pretty clearly inadequate, both on the ground that it is so deficient in economy as to be no account at all, and also because it turns the logical relations among judgments into *ad hoc* restrictions. The implication between judging conjunctively *A*,*B* and affirming *A* cannot be viewed as having its roots in the logical relations between the conjunction of *A* and *B*, and *A*.[21] The second alternative is more appealing. It is to express judgments in a modified Polish notation, so that the act becomes complex, but quite separable from the collected contents. The logical connections are then viewed as ultimately among acts of the mind and are not *ad hoc*, since these acts are built up in regular ways from specified basic acts: denying, conjoining, conditionalizing, disjoining, etc. This avoids the *ad hoc* nature of the first alternative and is lent some plausibility as well by some recent views on the nature of the structure and composition of sentences in natural languages, according to which many important structural features of the sentence are first composed in initial position and later spread throughout the sentence by transformations.[22]

As far as quantifiers are concerned, this alternative would treat prenex normal form as the ultimate normal form of judgment, so that the quantificational antecedents of all bound pronouns in a judgment would be fixed as the initial parts of the judging attitude. This prefix would be succeeded by the sequence of truth-functions, all of which precede the content. Thus the act of judging would be ultimately composed of certain atomic judgmental acts, and this molecular act would function upon a content, which would be an ordered sequence of atomic contents.

This view requires that atomic contents have in addition to sentential forms also predicative forms, but it seems in any case essential to allow such forms in order to account for *de re* beliefs.

The emphasis here on the prenex normal form for judgment in this form of the mentalistic view should not be taken as entailing that according to it the act precedes the content temporally. What is important for this view[23] is that we be able to give a description of the composition of the act as a complex whole, and that we look at the judgment as the application of this complex act to this complex content.

This mentalistic view does require the assumption that the mind has considerable structure and capacity beyond being a collection of contents. It is not apparent to what extent it requires the existence of an *agent* apart from the act. There is no obvious way to argue that an agent is required. A somewhat stretched use of the term *act*, following Hume and Russell, may make it seem that an agent is assumed, but this should not count as an argument.

This mentalistic view is not without its difficulties. One of these is one that arose also in the case of Hume's view; it is that there are problems in accounting on any mentalistic view for the dispositional sense of belief. Hume tried to account for this in the case of simple causal beliefs in terms of a causal regularity obtaining between the idea of the cause and that of the effect,[24] but this will clearly not do in the present case, since most judgments are not of this simple causal form, but consist rather in the application of a complex act to a complex content. It should be emphasized, however, that this is a difficulty for any mentalistic view and is thus no disadvantage for this theory as against other mentalistic accounts, but it is a difficulty nevertheless. In the present case a belief in the dispositional sense might be explained as a tendency to make a judgment consisting of acting upon a content of a certain sort, whenever the corresponding assumption is made.

## I.4. Behavioristic views. Pragmatism

Pragmatism is concerned not so much with an account of belief as with an account of true belief. It involves, nevertheless, a view of belief according to which the *operative character* of a belief is the class of actions to which it does or would lead. A belief is said to be *true* (in the pragmatic sense) if its operative character leads to long-run satisfaction.[25]

There are several ambiguities in this characterization: It makes a considerable difference whether we take operative character in the manifest sense, as consisting of the actions to which the belief does lead, or in the dispositional sense, as consisting of the actions to which the belief would lead in certain situations. The manifest sense raises the difficulties first, that many if not most of our beliefs have either no apparent manifest character or else have manifest characters which are quite uncharacteristic of the beliefs, due to the nature of the situations in which the belief happens to be implemented. A belief may lead to almost any action whatever, in combination with other beliefs, given appropriate circumstances. For this reason, the manifest sense of operative character seems to miss an important part of the motivation of the pragmatic account, namely to associate a belief with a class of characteristic actions. The spirit of that account requires that operative character be taken in the dispositional sense.

This means that the operative character of a belief is regarded as a functional connection between situations in which the agent may be, and actions to which that belief will lead that agent in those situations. This is a complication of the original definition, but an appropriate one.

We should, as far as possible, avoid reference to other beliefs in the definition of operative character and of truth. We wish to be able to consider the question of the truth of a belief without bringing into question the truth of others.

As far as the definition of truth is concerned, another ambiguity remains: We should distinguish satisfaction of the agent from satisfaction of the agent's desires, for each of these may occur with or without the other. If we count beliefs as true so long as their operative character leads to satisfaction of the agent, we get counterintuitive results; for a belief may lead me to undertake actions which satisfy me in unexpected ways, and we should not want to count such beliefs as true. I want to see a certain film, for example, and, believing it to be showing at the local theater, I go there. Another film is showing instead which, however, I enjoy. My mistaken belief about which film was showing led to my satisfaction without satisfying my desires. We should not on that account call it pragmatically true. We should, rather, count beliefs as pragmatically true when their operative characters satisfy the agent's desires, independently of whether the agent is satisfied or not.

A similar argument leads us to see that we should also distinguish which desires are satisfied by the actions: A mistaken belief which leads to an action which satisfies some desire of mine, not the desire in view of which I undertook the action, should not be counted on that ground as true. This means that the definition of pragmatic truth must be complicated still more, to something like: A belief is pragmatically true if the actions to which it, in concert with certain desires, would lead, would in turn lead to the satisfaction of those desires.

Thus the operative character of a belief is a functional connection between situations and desires on the one hand and actions on the other. It is not enough that it should connect situations and actions: We know the operative character of a man's belief in $A$ if, given his desires and a situation we can say what action (if any) his belief in $A$ will on account of those desires lead him to take in that situation.

In view of this feature of operative character, that it has to do in any given case with a number of situation-desire pairs and actions, we should not want to require, in order for a belief to be pragmatically true, that in *every* case the actions to which it leads satisfy the relevant desires. This would be rare, for in general the external and unforeseen features of the world will effect the consequences of the actions so that in some cases the relevant desires are satisfied and in some cases not. This means that the pragmatic account of truth requires some way of averaging over the satisfaction of desires, and further, that we are led by it to a concept of truth which is not classificatory, but either comparative or quantitative. For suppose $s_1, \ldots$ are situations, $d_1, \ldots$ sets of desires. Then the operative character of a belief is a function $b$ which in the situation $s_i$ for the desires $d_j$ leads to the actions $b(s_i, d_j) = A_{ij}$. In general if $b, b'$ are operative characters, we shall have that for some $i$ and $j$.

$$A_{ij} = b(s_i, d_j)$$

will lead to the satisfaction of desires in $d_j$, while

$$A'_{ij} = b'(s_i, d_j)$$

will lead to their frustration; and for some $i$ and $j$ we shall have the converse. Thus the pragmatic account will lead to the comparison of beliefs according to the extent to which the actions to which they lead satisfy the relevant desires. So long as (a) in each case, $d_i$ is a finite set of desires (b) we consider only a finite number of desire-situation pairs (c) all desires are given equal weight; then the comparison of beliefs according to the extent to which they are pragmatically true will be fairly straightforward. Some complications will arise in comparing functions $b$, $b'$ which are defined for different sets of situation-desire pairs, but these can be taken care of by comparing the proportions of cases in which the desires are satisfied rather than the absolute number of such desires. In general, though, the theory given under constraints (a), (b), and (c) above will be quite weak, and applicable only in relatively few and quite artificial cases, and it seems fairly clear that the major interest of the pragmatic theory will be in an account which is unconstrained in these ways. Just such a theory, developed by Ramsey and refined by later writers, is the account of belief based on utility functions. An exposition of this theory is given in the following chapter.

## I.5. Other behavioristic views

Pragmatism is one behavioristic view. There are others. Some of these are flatly reductionistic; they take belief to be nothing but a parameter or character of behavior. Others are not completely reductionistic but may claim, for example, that behavior is the only evidence relevant to questions of what a man believes.[26]

Behavioristic theories probably originate in the intuition that the differences among men's beliefs are just differences in how they behave, that if a man believed differently than he does, then he would behave differently than he does. One attempts then to develop a behavioristic theory in such a way that different beliefs correspond to different behaviors, so that the theory will enable us to give for each attribution of a belief to an agent, what the behavior is which the agent will undertake if and only if that attribution is true, if and only if that agent has that belief. As with the pragmatic account, with behavioristic theories generally we think of belief dispositionally rather than episodically, and this is perhaps the most significant difference from mentalistic views which account best for the episodic sense of belief and encounter characteristic difficulties in describing the dispositional senses of belief. From a behavioristic point of view we think naturally of belief as being enduring and dispositional. This means that, since a man has at a given instant many behavioral dispositions, he will, on a behavioristic account, have simultaneously many beliefs. For most of the plausible ways in which beliefs are correlated with sets of dispositions, beliefs which are consistent with each other may be correlated with conflicting sets of dispositions. Consider, for example, a pair of general beliefs of universal hypothetical form with consistent antecedents and inconsistent consequents.

(i) Packaged mushrooms are edible.
(ii) Yellow mushrooms are poisonous.

and suppose these to correspond to the dispositions

(iii) If he's hungry and encounters a packaged mushroom, he'll eat it.
(iv) If he's hungry and encounters a yellow mushroom, he won't eat it.

These dispositions conflict if the agent when hungry encounters a packaged yellow mushroom. Of course, the set of beliefs consisting of (i), (ii) and

(v) There are packaged yellow mushrooms.

is inconsistent, and thus upon encountering a packaged yellow mushroom, the agent has an inconsistent set of beliefs. But if we, the observers, know (v) to be true, and know also that the agent disbelieves it, then our account of his beliefs (i) and (ii) in terms of the dispositions (iii) and (iv) can't be quite right, since we attribute to him dispositions not both of which can be his. We think first of these dispositions as if they were causal pathways in the subject, each pathway always ready to be activated, rather like reflex arcs. But brief reflection shows that this can't be right, for it is easy, as in the above example, to think of situations to which the subject must, on these assumptions, respond inconsistently; although his beliefs are apparently consistent (though one of them is false), we describe a possible situation which leads the theory to an inconsistent assertion. The inconsistency in such a case is not in his beliefs – unless we wish to identify falsity and inconsistency – nor, of course, in his behavior; it is in the description which our theory issues as a prediction of what his behavior will be.

There are two apparent ways of meeting this difficulty: One of these amounts to supposing that people's beliefs are in general not so simple as (i) and (ii). The second way is to drop the identification of specific beliefs with specific dispositions. In both of these we should avoid attributing dispositions

In situation $A_1$ he will do $B_1$.
In situation $A_2$ he will do $B_2$.

to the subject if $A_1$ and $A_2$ are consistent and $B_1$ and $B_2$ are inconsistent and attribute instead.

In situation $A_1$ and not-$A_2$ he will do $B_1$.
In situation not-$A_1$ and $A_2$ he will do $B_2$.
In situation $A_1$ and $A_2$ he will do $C$.

where $C$ is some consistent description. The eventuation will be a set of dispositions such that every one of its subsets, the antecedents of which are consistent, will have a consistent set of consequents.

Given that beliefs are always characterized by such sets we can no longer hold both that people have such simple beliefs as we ordinarily attribute to them and that having a belief amounts to having a specifiable disposition. For on such a list of dispositions there will be no simple dispositions to correlate with simple beliefs. We may say that every belief is correlated with (or perhaps just *is*) a unique disposition, but then the beliefs will be quite complicated and we shall be able to make little sense out of ordinary simple belief attributions. Or, on the other hand, we may avoid unique correlation or identification of beliefs with dispositions and instead view the manifestations of belief in behavior as diffuse, being spread out throughout the complex dispositions. In this latter case, clearly, we can not hold on to any reductionist claims, at least not of the original straightforward sort, i.e., we cannot claim that attributing a belief to a man amounts to attributing a behavioral disposition to him, for there will in general be no specifiable such disposition.

Of these two modifications the more plausible, at least on the face of it, is the holistic alternative, in which the totality of an agent's beliefs is correlated with the totality of his behavioral dispositions. This permits the beliefs to be simple, specifiable at least approximately as we ordinarily take them to be, while the behavioral dispositions are, as we should expect, quite complex. There are at least the following two important consequences of this approach:

(i) One of the original motivations of the behavioristic view is abandoned, we can no longer hold that to attribute a belief to a man is to attribute a way of behaving to him.

(ii) The correlation of the totality of behavioral dispositions will require in general that we take account also of the agent's desires, since one thinks of action as stemming from belief and desire together, not from either alone. As soon as this move is made, it is practically inevitable that we think of desires as well as of beliefs as varying in strength, since we shall view the action as the net result of forces which are only partially in conflict. This requires that the forces be comparable in strength. Thus behavioristic views in general, as did the pragmatic theory in particular, lead us to the consideration of partial belief.

## NOTES

[1] Cf. *Scheffler*.
[2] Cf. *Hume*, pp. 132, 135.
[3] Cf.; for example, *Russell* (1903) section 48.
[4] *Kaplan*.
[5] *Quine* (1960), *Church*.
[6] *Hume*, throughout Book I.
[7] Cf. Section 2 of Tarski 'The Concept of Truth'.
[8] Cf. Ramsey 'General Propositions'.
[9] 'The Ideas of Space and Time' in *Hume*.
[10] *Hume*, p. 94.
[11] *Whitehead and Russell*, pp. 174f.
[12] Cf. *Brentano*, Chapter VII.
[13] *Hume*, pp. 180f.
[14] *Meinong*.
[15] "They are the successive perceptions only, that constitute the mind", *Hume*, p. 253.
[16] See for example, 'On the Nature of Truth and Falsehood'.
[17] *Meinong*, and *Russell* (1904).
[18] Cf. *Ramsey*, p. 146.
[19] In *Russell* 'On Propositions', p. 308.
[20] "I incline to the view that the difference [between a memory and an expectation] is not in the content of what is believed, but in the believing...[D]ifference of tense, in its psychologically earliest form, is no part of what is believed, but only of the way of believing it; the putting of the tense into the content is a result of later reflection".
[21] Cf. *Ramsey*, pp. 146ff.
[22] Cf. *Chomsky*, Chapter 3.
[23] *Chomsky*, pp. 139f.
[24] *Hume*, p. 170.
[25] Cf. *Russell*, 'On The Nature of Truth and Falsehood'.
[26] Cf. *Carnap*, (1959). *Ryle* explicitly holds a behavioristic view of emotion and probably would have given a similar view of belief had he considered the issue.

CHAPTER II

# PARTIAL BELIEF

## II.1. MENTALISTIC PARTIAL BELIEF

Hume's account of partial belief is an extension in a quite natural way of his account of non-partial belief.[1] Partial belief is a consequence, on his view, of the mind's capacity to divide its force equally among distinct alternatives. This is most simply applicable to *conditional* partial beliefs, of the form '$B$ will be consequent upon $A$'. The strength of such a belief will, he says, be $m/k$ just when:

(i) The lively conception (belief in $A$) is followed in the mind by the lively conceptions of some distinct $C_1, \ldots, C_k$.

(ii) Just $m$ of these cases are seen to be $B$.

Hume's account assumes the principle of indifference. The alternatives in which $B$ obtains are supposed to contribute equally to the strength of the belief in $B$.

It is quite clear that this account is intended to be probabilistic, that is to conform to the laws.

(iii) If $B$ is a necessary consequence of $A$, then the strength of belief in $B$ given $A$ is 1.

(iv) If $B_1$ and $B_2$ are incompatible in the presence of $A$, then (assuming conditional belief is defined in $B_1$ given $A$ and in $B_2$ given $A$) the strength of belief in $B_1$ or $B_2$ given $A$ is the sum of the strength of the beliefs in $B_1$ given $A$ and in $B_2$ given $A$.

The second of these offers a difficulty of the same sort which arose in the case of negations in non-partial beliefs: If $B_1$ and $B_2$ are believed incompatible, then, in Hume's view, they can never be conceived together. Thus there is no way in which the conception of $A$ can be followed by the simultaneous conception of the incompatible $B_1$ and $B_2$. This difficulty seems endemic, and can be avoided only either by providing some other account of incompatibility, or by supposing the mind to be capable of simultaneous distinct incompatible ideas. The first of these brings up the problems in the Humean account, or lack of an account, of negation. These problems would result if, for example, incompatibility were taken to be a substantive part of the judgment of $B_1$ and $B_2$'s incompatibility. (If this judgment were thought of as conceiving of incompatibility, of $B_1$, and of $B_2$.) The second amounts to abandoning the view that there is but one act of the mind.

Indeed, all of the problems in Hume's account of non-partial belief arise anew for his account of partial belief. It is, for example, not easy to see how to apply the law that the sum of probabilities of $B$ given $A$ and of not-$B$ given $A$ should be 1, for the same reason that beliefs in negations are hard to explain on the lively conception view. Similar problems arise, of course, with other logical notions and are particularly critical with quantifiers: Whereas in the case of full belief it is at least consistent, if not quite plausible, to hold that believing that something is $F$ means believing, of something, that it is $F$, it is clearly wrong to hold that the strength of belief in $\exists x \cdot Fx$ given $A$ is just, for some $a$, the strength of belief in $Fa$ given $A$. Indeed, we should probably want to maintain that if $\mathfrak{D}$ is the set of all individuals for which belief is defined, then belief in $\exists x \cdot Fx$ given $A$ is no less than the sum, for $a$ in $\mathfrak{D}$, of beliefs in the various $Fa$ given $A$, so long as $A$ involves no $a$ in $\mathfrak{D}$.

Another difficulty which arises with Hume's account is an instance of the difficulty of embedding in the content of judgment, features of judgment which are analyzed as part of the act. (Cf. I.2, above.) The problem arose in the discussion of non-partial belief in connection with *reductio* argument. Now, in the case of partial belief, it comes up in attempting to account for probability judgments. What status, for example, is to be given to the multiplication law?[2]

(v) $P_{AC}(B) \cdot P_A(C) = P_A(ABC)$.

And, should one insist upon interpreting this as a psychological law, there are still the problems of embedding probability judgments, for example;

(vi) For none of the $F$'s is the chance of it being $G$ greater than $\frac{1}{2}$.

(vii) Either the chance of $B$ given $A$ is $m$, or the chance of $C$ given $A$ is $k$.

One way of responding to this difficulty is to claim that the account of partial belief does not pretend to be an interpretation of *all* statements of the probability calculus, or of all probability judgments; that the intention is to provide an analysis of the phenomena of partial belief. The believing is analyzed, in this account, as part of the act, and the strength of believing, in terms of the apportionment of this force over the parts of the content or contents. There are, of course, other uses of the probability calculus which may not be comprehensible in this way.

It is not easy to separate, in Hume's account of partial belief, the problems which arise from the definition of partial belief in terms of shared force, from those which arise from the assumption of only one act of the mind. Hume views the account of partial belief as the comprehensive account, and he thinks of non-partial belief as a special case of partial belief when all the force is directed upon one object.

If we try to allow, say, *assumption*, as well as judgment as an act of the mind, we see that it would not be easy to allow partial assumption, as distinct from partial belief. This is because it is the force of *conviction*, a force proper to judgment and which distinguishes it from assumption, which is shared by the several consequents in partial belief: Judging that *B* is consequent upon *A* with a certain strength is, in the Humean account, spreading the force of conviction consequent upon judging *A* over contents, in a certain number of which *B* obtains. Since there is no 'force of assumption' in the case of assuming, there can be no parallel account of *assuming* that *B* is consequent upon *A* with a certain strength. This is because, in the judgment that *B* is consequent upon *A* with a certain strength the content consists of *A* and *B*, in that order. The strength of the consequence relation is analyzed as part of the content. In the case of, for example, truth functions, we were led to give a parallel structure to judging and assuming, so that there are conjunctive judgments and conjunctive assumptions, negative judgments and negative assumptions, and so on. The important difference between that account and the present case is that none of the truth functions were analyzed in terms of strength of conviction. For this reason it was possible to give in the case of truth functions parallel accounts of judgment and assumption, the parallelism consisting in (i) the isomorphism of structure between a judgment and the corresponding assumption. (ii) the identity of the object, the truth conditions, for corresponding judgments and assumptions. It is not, however, possible to construct such parallel accounts in the case of partial belief, since there is no feature of assumption to correspond to the force of conviction present in judgment upon which the account of partial judgment is based.

One might, of course, think of describing the assumption that *B* is consequent upon *A* with a certain strength as an assumption about a judgment, of course such second order assumptions (and corresponding judgments) are possible, but this assumption would not be an assumption with the same content as the judgment that *B* is consequent upon *A* with that strength. The content of the latter consists of *A* and *B*. The content of the second order assumption involves also reference to judgment.

Other analyses of probability may provide an account of first order judgment and assumption in which the probability is analyzed as part of the content, but I think it clear that Hume's account cannot do this. This raises the question of what relation there can be between partial belief and probability judgments in which probability is a part of the content.

It is not open to us to treat probability as interpreted in partial belief as we did the logical connectives; by introducing a parallel structure of assumptions, like judgments, but lacking conviction. This does not mean, however, that a theory of assumptions or other mental acts will have no place in a mentalistic account of partial belief. For whatever may serve as the content of a non-partial belief may also, on the mentalistic account, serve as the content of a partial belief, and thus we should not want to identify force of conviction with liveliness of conviction, for reasons similar to those which prevent this in the case of non-partial belief: It would for example, mean that whoever judged not-$B$ to be consequent upon $A$ with a certain strength would also judge $B$ to be consequent upon $A$ with that strength.

## A. TRANSPARENCY OF PARTIAL BELIEF AND DE RE MENTALISTIC BELIEF

The way in which Hume's account of partial belief is, or approaches being, probabilistic, depends upon notions of necessity and incompatibility; they are essentially involved in the laws (iii) and (iv) on page 39 above. Hume thought of these notions as defined in terms of *conceivability*,[3] $B$ being a necessary consequence of $A$ if $A$ and not-$B$ is inconceivable, and $B_1$ and $B_2$ being incompatible given $A$ if $A$ and $B_1$ and $B_2$ is conceivable. This notion of conceivability gives rise to an equivalence relation, according to which $A$ and $B$ are equivalent if both ($A$ and not-$B$) and ($B$ and not-$A$) are inconceivable, and it is clear that Hume's notion of partial belief is *transparent* for this equivalence relation, in the sense that strength of belief is invariant under replacement of equivalents. Further, we can also see that this is the strongest notion of transparency under which such invariance is guaranteed: For suppose that $A$ and $B$ are not equivalent in this sense. We can assume without loss of generality that $A$ and not-$B$ is conceivable. Then judging $A$ may be succeeded by the lively conception of alternatives in not all of which $B$ is seen to obtain, so the strength of belief in $B$ given $A$ may be less than 1. But the strength of belief in $A$ given $A$ is 1. Thus if $A$ and $B$ are not equivalent, invariance of strength of belief under exchanges of $A$ and $B$ is not guaranteed.

In this argument the extent of transparency of partial belief is related to the probabilistic nature of partial belief. The relation is that the transformations under which partial belief is transparent are precisely those which give the concept of *necessity* for which partial belief is probabilistic. We shall return to a generalized form of this argument below.

Mentalistic accounts of partial belief characteristically discuss belief in individual objects or in propositional objects. The *de re* senses of belief, where a given object is believed to be of a certain sort, are not usually accounted for by them. In Hume's theory the reason for this is that he is attempting to develop a theory of belief in terms of mental objects; there is no way in his view to refer to an object except by referring to an impression or an idea of it. Thus it is difficult in his theory even to express a *de re* belief, much less to give an account of such.

There is an important way in mentalistic theories in which we can consider *de re* beliefs. This is by taking the objects of belief to be properties. Then the strength of partial belief relating two properties will be the strength of the preserved evidential relationships between the properties. Hume's nominalism prevented him from characterizing belief in properties, he took the objects of belief as much as possible to be particulars. There are however, some good reasons for defining partial belief in properties: For one thing, it is intuitively plausible that we have such partial beliefs. We take the question 'What are the chances that an $F$ is also a $G$?' to be sensible. We also take such beliefs to be probabilistic. Secondly, these relations function in an important way in statistical inference; where the inferential relations concern properties to a far greater extent than they do individuals. Further, the mentalistic account is also easily applied to conditional beliefs concerning properties: My belief in $F$ conditional upon $G$ is $m/k$ if the lively conception of $F$ is succeeded by the lively conception(s) of the properties $H_1, \ldots, H_k m$ of which are seen to entail $G$.

## II.2. THE RELATION OF BELIEF AND DESIRE

In Thomas Bayes' posthumous paper of 1763 he defined the probability of an event to be

> ... the ratio between the value at which an expectation depending on the happening ought to be computed, and the value of the thing expected upon its happening.[4]

It is quite within the spirit of the time and of Bayes' paper, and makes good sense of the definition as well, to take this to define strength of belief. The definition requires, what is not provided by Bayes, some metrical account of *value*, so as to be able to express numerically the values of expectations and values expected. Bayes' paper just assumes such an account and that the measurement of value which it develops will uniquely determine the ratios of the definition. It is important to remark at the outset that this may be accomplished without developing a method of measurement which assigns a unique numerical quantity to each thing and expectation. If belief in an event is the ratio

$$\frac{\text{value of the expectation depending upon the event}}{\text{value of the thing expected}}$$

then in order for belief to be uniquely determined numerically, this ratio must be uniquely determined, which allows that there may be different scales for measuring values, so long as this *ratio* has the same numerical value in all scales used to determine the values in its numerator and denominator, the same scale, of course, being applied to both. The measurement of *weight* provides a useful example:[5] The quantity

$$\text{weight of } a$$

is not uniquely determined, since it may be given in pounds, kilograms, or some other scale. No matter which of these scales is used, however, the ratio

$$\frac{\text{weight of } a}{\text{weight of } b}$$

is, for given $a$ and $b$, a constant. This is consequent upon the simple relations which hold among different scales for the measurement of weight: Take $W_1$ and $W_2$ to be ways of measuring weight, and let $W_1(a)$ and $W_2(a)$ be the numerical weights of the object $a$ as assigned by $W_1$ and $W_2$ respectively. Any two such scales of weight are related by multiplication by a positive constant: That is to say, there is some positive constant $m$, such that for every $a$

$$W_2(a) = m \cdot W_1(a).$$

For example,

$$\text{kilograms } (a) = 2.2046 \cdot \text{pounds } (a)$$
$$\text{pounds } (a) = 0.4536 \cdot \text{kilograms } (a)$$

for every object $a$. Thus the ratios

$$\frac{\text{pounds }(a)}{\text{pounds }(b)} \qquad \frac{\text{kilograms }(a)}{\text{kilograms }(b)}$$

are equal for all objects $a$ and $b$, and thus the ratio

$$\frac{\text{weight of } a}{\text{weight of } b}$$

is determined uniquely by the objects $a$ and $b$, independently of what scale of measurement is used.

### A. MEASUREMENT AND INVARIANCE

This characteristic of measurement of weight may be thought of as consequent upon the arbitrariness of the *unit* of weight, and the non-arbitrariness of the meaning of *zero weight*. It is clear that what is weightless should have the same weight, zero, in all scales of weight, and thus, as it is sometimes said, that there is a *natural zero* for weight, but there is no apparent natural unit of weight, and thus different scales may be related in terms of the translations of their units. The distinction between what is fixed and what is arbitrary in the measurement of weight is founded in the empirical system of an equal-arm balance which is thought of as giving the meaning of that measurement. That system fixes a zero, but leaves the unit quite undetermined.

One more example will serve us well for the exposition of the measurement of value. That is the measurement of *temperature*. Ignoring the absolute measurement of temperature, temperature differs from weight in lacking not only a natural unit, but a natural zero as well. As a consequence of this, the translation of temperature from one scale to another requires multiplication, to accommodate the difference in unit, and also the addition of a constant, to translate the zero point. Thus, for example,

$$\text{fahrenheit temperature }(a) = \tfrac{9}{5}\cdot\text{centigrade temperature }(a) + 32.$$

More generally, if $T_1$ and $T_2$ are two scales of temperature, there are a positive real number $m$ and a real number $n$, such that for each object $a$

$$\text{(i)} \qquad T_2(a) = m\cdot T_1(a) + n.$$

Consequently, the ratios of temperatures are not given independently of scale. That is to say, we may have

$$\frac{T_1(a)}{T_1(b)} \neq \frac{T_2(a)}{T_2(b)}$$

for given objects $a$, $b$, and scales $T_1$, $T_2$. In order to find an invariance of ratios, we must consider not simply temperatures, but the ratios of their *differences*. It is a consequence of (i) above that the ratio

$$\frac{T(a) - T(b)}{T(c) - T(d)}$$

is invariant for given objects $a$, $b$, $c$, and $d$, no matter which scale of temperature is used to measure it. It is a further important consequence that the *average* of two temperatures is also invariant in the following sense: If the temperature of the object $a$ is midway between those of $b$ and $c$, that will be so no matter which scale is employed to measure temperature:

$$T(a) = \frac{T(b) - T(c)}{2}$$

is either true for all scales, $T$, of temperature or true for none.

## B. MEASUREMENT, VALUE AND PREFERENCE

Bayes' definition of probability, understood as a definition of strength of belief, says, roughly, that a man's belief in a proposition is the ratio of the greatest amount he will risk to gain a stake if the proposition be true, to the size of the stake. Belief in $A$ is just $p/s$ if the believer will put up $p$, on condition that he is to receive $s$ if $A$ occurs and nothing if $A$ does not occur. This definition requires for its consistent application that there be some way of measuring value which leads, as does the measurement of weight, to the invariance of ratios of values, independently of the particular scale employed. One thinks immediately of monetary worth as providing just such a standard: There is a natural zero; no money at all; and different currencies are always related by multiplication by a positive constant. In addition to the limited applicability of this, however, there being for many people propositions which have no clear monetary value, there is also the difficulty that the odds at which one will bet are in general strongly affected by the size of the stake. A man of moderate means may be willing to pay \$100 yearly to insure himself against a \$3,000 loss in the event of an accident to his automobile, but the same man with the same means would pay, let us say, no more than \$40 to insure himself against a \$1,500 loss. This does not mean that he has computed badly, or that he suffers from inconsistency, but only that one loss of \$1,500 would not affect him as severely as would a second \$1,500 loss given that he has lost that much already.

This means that monetary stakes and values cannot be used in the application of Bayes' definition.

The reaction to this difficulty has been for the most part to attempt a definition of *value* in terms of preference which, first, has the requisite invariance characteristics, and, second, is not subject to ratio variations with the quantities of values involved. That is to say, such that if $p$ is the most a man will put up to receive $s$ if $A$, and $p'$ is the most he will put up to receive $s'$ if $A$, then

$$p/s = p'/s'.$$

The requisite invariance characteristic, that

$$\frac{v(a)}{v(b)} = \frac{v'(a)}{v'(b)}$$

if $v$ and $v'$ are scales for measuring value, causes no difficulty in the monetary measurement of value, since that measure has an obvious natural zero, and different currencies may always be made to correspond by a constant multiplication. When we turn, however, to the computation of preference between events or propositions, we find there no such natural zero. This is not to say that no such point can be fixed in a good way,[6] but only that there is no apparent point of absence of value, as there are points of weightlessness and the absence of wealth. This is an indication, to be supported below, that the requisite invariance of ratios is not forthcoming. In fact, what we get in the standard technique is a method of assigning values to propositions which is like temperature in that if $v$ and $v'$ are two scales which conform to the method, then there are constants $m$ and $n$, $m>0$, such that for all $a$[7]

$$v'(a) = m \cdot v(a) + n.$$

The consequence is that the ratios

$$\frac{v(a)}{v(b)}$$

will not be invariant for different scales $v$. Ramsey,[8] in particular, met this problem by focussing on the invariance that *is* guaranteed, namely that of

$$\frac{v(a) - v(b)}{v(c) - v(d)}$$

and of the equivalence of

$$v(a) = \frac{v(b) + v(c)}{2}$$

for all scales $v$.

## C. RAMSEY'S METHOD

To exploit these features Ramsey conceived of a way to place bets in terms not of values simply, but in terms of value intervals.

Suppose the quantities of value $x$, $y$, and $z$ are such that $y > x > z$, and that a man is indifferent between (i) $x$ for certain, and (ii) a wager which gives him $y$ if $A$ and $z$ if not-$A$. This man is guaranteed at least $z$ units of value, and we may think of him as willing to risk the increment $(x-z)$ on condition that he receive the increment $(y-z)$ should $A$ occur. That is to say, the value of the expectation depending upon the happening of $A$ is the increment $(x-z)$, and the value expected upon the happening of $A$ is $(y-z)$. Thus, by Bayes' definition, the man's belief in $A$ is the ratio

$$\frac{x - z}{y - z}.$$

This ratio is invariant when $x$, $y$, and $z$ are all transformed by an operation

$$m(\xi) + n.$$

This led Ramsey, after developing an appropriate measurement of value, to define strength of belief in a proposition $A$ as the ratio

$$\frac{v(a) - v(c)}{v(b) - v(c)}$$

where the subject is indifferent between the proposition $a$, and a wager which yields the proposition $b$ if $A$, $c$ if not-$A$.

## D. AN AXIOMATIC DESCRIPTION

The measurement of value, developed by Ramsey and refined by others, which supports this definition, starts with the assumption of a collection of propositions, among which preference and indifference are defined for a given subject. Preference is assumed to order the collection transitively and to be asymmetric. Indifference is assumed to be an equivalence relation, and propositions neither of which is preferred to the other are supposed indifferent. Writing $\succ$ for preference and $\sim$ for indifference, these assumptions may be put axiomatically:[9]

AXIOM 1. *There is a collection* $\mathbb{Q}$ *of propositions in which the relation* $\succ$ *of preference and the relation* $\sim$ *of indifference are defined.* $\succ$ *is transitive and asymmetric.* $\sim$ *is an equivalence relation, and* $x \sim y \Leftrightarrow$ *neither* $x \succ y$ *nor* $y \succ x$. $\mathbb{Q}$ *is not null, and includes at least two propositions, one of which is preferred to the other.*

Axiom 1 alone permits the assignment of quantities of value to the members of $\mathbb{Q}$ conforming to preference and indifference in the sense that

$$v(a) > v(b) \Leftrightarrow a \succ b$$
$$v(a) = v(b) \Leftrightarrow a \sim b$$

these principles remaining invariant, for given $a$ and $b$, for different scales $v$ of values. A1 is not, however, sufficient to permit measurement for which the ratios of value intervals are invariant. For this some interpretation must be provided for *preference intervals* or *increments*. Ramsey's technique for doing this is as follows: There are first found propositions $a$, $b$ and $E$ such that

(i) $a \succ b$

(ii) The agent's preferences for $a$ and $b$ are unaffected by the presence or absence of $E$. (He is indifferent, for example, between $a$ in conjunction with $E$ and $a$ without $E$.)

(iii) The following two wagers are indifferent

$a$ if $E$, $b$ if not-$E$

$b$ if $E$, $a$ if not-$E$.

The wagers in (iii) are then said to mark the *preference midpoint* between $a$ and $b$. It is assumed that if $a$ and $b$ are any non-indifferent propositions, some $E$ can be found satisfying (ii) and (iii), and thus that each pair of non-indifferent propositions has a preference midpoint. These midpoints are taken to be in the domain of the subject's preference and indifference relation. It is also supposed that the preference midpoint of indifferent propositions is indifferent to both of them. These and some other important assumptions may be made explicit in the form of an axiom. We write

$$m(a,b)$$

for the preference midpoint of $a$ and $b$.

AXIOM 2. *The collection* $\mathbb{Q}$ *is also closed under the formation of preference midpoints; if a and b are in* $\mathbb{Q}$ *then* $m(a,b) \varepsilon \mathbb{Q}$. *Further,*

(1) $m(a,b) \sim m(b,a)$;
(2) $a \sim b \Leftrightarrow a \sim m(a,b)$;
(3) $a \succ b \Leftrightarrow a \succ m(a,b) \succ b$;
(4) $m[m(a,b), m(c,d)] \sim m[m(a,c), m(b,d)]$;
(5) $a \succ b$ and $c \succ d \Rightarrow m(a,c) \succ m(b,d)$;
(6) $a \sim b$ and $c \sim d \Rightarrow m(a,c) \sim m(b,d)$.

Axioms 1 and 2 may be seen to be conjunctively consistent by taking $\mathbb{Q}$ to consist of the non-negative rational numbers in some interval,[10] interpreting $\succ$ and $\sim$ as $>$ and $=$, and $m$ as giving the average of its arguments. Hence, if $\mathbb{Q}$ were a collection of propositions conforming to Axioms 1 and 2, it could be made to correspond with some interval $\mathbb{Q}\#$ of rational numbers in such a way that, where $a^{\#}$ is the number corresponding to the proposition $a$,

$$a \succ b \Leftrightarrow a^{\#} > b^{\#}$$
$$a \sim b \Leftrightarrow a^{\#} = b^{\#}$$
$$m(a,b)^{\#} = \tfrac{1}{2}(a^{\#} + b^{\#}).$$

The following axiom is simple and very convenient, but may not be plausible as regards preference midpoints and propositions. It can be dispensed with in favor of less controversial principles, but it is included here for expository convenience.

AXIOM 3. *For any propositions a and b there is some proposition c such that* $b \sim m(a,c)$

It remains to support the quantitative comparison of intervals of preference or value. The end of this will be to characterize a relation

$$abT(m/n)c$$

which will say of propositions $a$, $b$ and $c$ (assume for now that both $b$ and $c$ are preferred to $a$) that the ratio of the interval between $b$ and $c$ to that between $a$ and $c$ is $m/n$. This will be understood to mean that there are propositions $b_2, \ldots, b_n$ and $c_2, \ldots, c_m$, which are ordered preferentially somewhat like this

$$\begin{array}{cccccc} a & b & b_2 & \ldots & b_{n-1} & b_n \\ a & c & c_2 & \ldots & c_{m-1} & c_m \end{array}$$

where the intervals of preference are the same within each row. Given this, the choice of propositions $a$ and $b$, with $b \succ a$, will correspond to picking propositions of value zero and one, and the value of a proposition $c$, preferred to $a$, will be just $m/n$ in the scale determined by $a$ and $b$, when $abT(m/n)c$.

The characterization of the relation $T$ depends upon a simple recursively defined relation $M^n$.

$$\begin{cases} abM^1cd \Leftrightarrow a \sim c \quad \text{and} \quad b \sim d \quad \text{and} \quad b \succ a \\ abM^{n+1}cd \Leftrightarrow. \text{ For some } e, abM^n ec \quad \text{and} \quad c \sim m(e,d). \end{cases}$$

Thus, in the above diagram,

$$abM^n b_{n-1} b_n$$
$$acM^m c_{m-1} c_m.$$

The commensurability in this way of every appropriately situated triple is assured by

AXIOM 4. *If* $b \succ a$ *and* $c \succ a$ *then there are positive integers* $m$ *and* $n$ *and propositions* $b'$, $c'$, $b''$, $c''$, *such that*

$$abM^n b'b'' \qquad acM^m c'c''.$$

The plausibility of these axioms as descriptive or prescriptive of preference is easily questioned. Axiom 1 entails that every pair of propositions is comparable with respect to preference, and that is pretty clearly not in general true. It is also implied that any finite chain of propositions, each of which is indifferent to the next, consists entirely of indifferent propositions; but it may be presumed possible in many cases to connect a proposition to a preferred proposition by a chain of such a sort, the distinctions between adjacent members being below the threshold of notice. If we think antecedently of values as given by a numerical assignment, then it is plausible that such an assignment satisfy Axiom 2. But the relation of numbers and values is just what is at issue in these axioms, and they should be considered, therefore, without such a presumption. Thus, the density characteristic assumed in (3) of axiom 2 raises problems of the distinguishability of valued alternatives which may be made arbitrarily proximate in value.

The difficulty with Axiom 3 is obvious. It says that any preference interval may be reiterated indefinitely often. Thus, no matter how dreadful and desirable may be options $a$ and $b$, $a$ stands midway between $b$ and some even more dreaded alternative, and there is some proposition as much preferred to *a* as it is to *b*.

But without denying any of these arguments, or any of the more general arguments offered later, it should nevertheless be remarked that it is nothing short of astounding that a set of principles as plausible as these were found which actually permit the measurement of value in a quite rigorously determined way.

## E. SOME THEOREMS

The way in which these assumptions permit the measurement of value is revealed in a sequence of theorems stated here without proof.[11]

THEOREM 1. If $\mathfrak{Q}$ is a set of objects on which relations $\succ$ and $\sim$ and a function $m$ satisfying the axioms are defined, then $\mathfrak{Q}$ may be partitioned into equivalence classes or *cosets* by the relation

$$\alpha_a = \{b \mid a \sim b\}.$$

The relations $\succ$ and $\sim$ among members of $\mathfrak{Q}$ will then induce relations $\succ^*$ and $\sim^*$ among these cosets, and the function $m$ will induce a function $m^*$. We shall have

$$\begin{aligned} &\alpha_a \succ^* \alpha_b \Leftrightarrow a \succ b \\ &\alpha_a \sim^* \alpha_b \Leftrightarrow a \sim b \\ &m^*(\alpha_a,\alpha_b) = [m(a,b)]^*. \end{aligned}$$

The relations and functions so induced will conform to Axioms 1 through 4 in the family $\mathfrak{Q}^*$ of cosets. $\sim^*$ will be just the identity on $\mathfrak{Q}^*$, and $\succ^*$ will be transitive, asymmetric and connected in $\mathfrak{Q}^*$.

THEOREM 2. Given a collection $\mathfrak{Q}$ conforming to Axioms 1 through 4, and a corresponding $\mathfrak{Q}^*$ as given in the preceding theorem, $\mathfrak{Q}^*$ may be homeomorphically mapped into the rational numbers by a function $\varphi$ which assigns some rational number to each coset in $\mathfrak{Q}^*$ in such a way that, for $\alpha$, $\beta$, $\gamma$ and $\delta$ members of $\mathfrak{Q}^*$.

$$\begin{aligned} &\alpha >^* \beta \Leftrightarrow \varphi(\alpha) > \varphi(\alpha) \\ &\alpha \sim^* \beta \Leftrightarrow \alpha = \beta \Leftrightarrow \varphi(\alpha) = \varphi(\beta) \\ &\varphi[m^*(\alpha,\beta)] = \tfrac{1}{2}(\varphi(\alpha) + \varphi(\beta)) \\ &\alpha,\beta M^n\, \gamma, \delta \Leftrightarrow \varphi(\beta) > \varphi(\alpha) \quad \text{and} \quad \varphi(\beta) - \varphi(\alpha) = \varphi(\delta) - \varphi(\gamma) \\ &\quad \text{and} \quad n[\varphi(\beta) - \varphi(\alpha)] = \varphi(\delta) - \varphi(\alpha) \end{aligned}$$

$\varphi$ will further support a homeomorphic mapping $\psi$ from $\mathbb{Q}$ into the rational numbers

$$\psi(a) = \varphi(\alpha_a).$$

THEOREM 3. (i) If $\psi_1$ and $\psi_2$ are two homeomorphisms from $\mathbb{Q}$ into the rational numbers as described in Theorem 2, then there are rational numbers $m$ and $n$, with $m > 0$, such that for all $a$

$$\psi_2(a) = m\psi_1(a) + n.$$

(ii) Further, if $\psi_1$ is a homeomorphism from $\mathbb{Q}$ into the rational numbers as described in Theorem 2, and if $\psi_2$ is defined as above for given $m$ and $n$ with $m > 0$, then $\psi_2$ is also such a homeomorphism.

## II.3. Difficulties with this Account

There are several difficulties with this account of belief, some more serious than others, and some of which may be met, at least in part, by modification of it.

### A. DIFFICULTIES OF LIMITED APPLICATION

The first of these applies quite generally to theories in which the belief in propositions is associated with desire for the anticipated consequences of acting upon them. The crudest claim of this sort is: If a man wants a certain reward and believes that a certain action on his part will bring him that reward, then he will do that action. Of course this is not right, because it makes no distinction between great and trivial rewards or between strong and weak beliefs, but these objections may temporarily be put aside. The further problem, which is generalizable to less crude claims and theories, is that the only beliefs considered directly are beliefs about actions of the believer. This is a difficulty shared by the practical syllogism[12] as an account of the relation of belief, desire and action: The action which concludes the syllogism must be mentioned in the premises, and hence the judgments or beliefs which lead to the action must include some proposition about the action in their content.

The Ramseyian account of partial belief shares this difficulty for, though the propositions for which the belief function is defined may themselves not include reference to actions, it is a clear assumption that the action of wagering is the action which leads to reward, appropriately weighted by degrees of belief and amounts of utility increment. The beliefs with which the theory is directly, if sometimes not explicitly, concerned are of the form 'Wagering on *A* will bring me *y* if *A* and *z* if not-*A*', and the relation of the belief in *A* to the agent's actions is intermediated by his beliefs about the effects of those actions.[13]

This seems wrong, first for the reasons that beliefs may apparently lead to actions without those actions being reflected upon, and second because it seems that the cognitive origins of actions are not restricted to beliefs about the effects of those actions. Whether or not one may act with no concern for effect, it seems clear that one may act in the light of other concerns as well. Beliefs about duties, memories, and beliefs about the pleasure or displeasure concomitant with anticipated actions (it is not, for example, plausible to think of the pleasures of sex as effects of sexual actions, or of the enjoyment of playing music as an effect of that playing) are examples of this.

In the case of non-partial beliefs a ramification of the crude theory which is not directly subject to these objections and which meets other difficulties as well, is to relate beliefs and desires as a body with actions totally considered, and not to attempt a correlation of specific beliefs with specific actions. A rough but serviceable formulation of this is: A man's beliefs are such that were they true his actions would satisfy his desires.

This formulation has the advantages that belief and desire are explicitly parameters of action, that the totality of actions is seen as determining the totality of beliefs and desires, and that it does not assign particular importance to beliefs about actions. It has as formulated the disadvantage or unclarity that beliefs and desires are not uniquely given by it, and in particular that every inconsistent set of propositions is believed by everyone, and that any desire which cannot possibly go unsatisfied is everyone's desire. These may not be conclusive difficulties, but they are not as trivial as they may at first appear, and to avoid or to overcome them seems not to be a matter of mere detail. What is of more interest here, however, is the incompleteness or the artificiality of the formulation; it makes no allowance for partial beliefs or desires of varying strength, and there is no apparent way of applying it to these. The obvious ways of modifying it to overcome this all seem to result in variations of the expected utility hypothesis; the claim that people perform those actions which they believe will maximize expected utility increments. And that hypothesis has just the difficulties mentioned at the beginning of this section.

### B. A SECOND SUCH DIFFICULTY

In order to measure your belief in a proposition by Ramsey's methods I must construct a gamble the stakes of which are other propositions which you are convinced that I can bring about, and I must find some third independent outcome which is between the other two stakes in utility. If $A$ is the proposition the belief in which is being measured, and $y$, $z$, and $x$ are the outcomes, then $x$ must be between $y$ and $z$ in preference. At least that is the situation when neither the proposition $A$ nor its denial enhances either $y$ or $z$. It is consistent with this scheme that $A$ be independently valuable, so long as its value is unaffected by the presence or absence of $y$ or $z$; if $A$ is valuable $y$ will serve as a bonus in the event of its occurrence and $z$ as a recompense in the event of its non-occurrence. That is an advantage of Ramsey's method,[14] that it does not presume an exclusive division of propositions into those which are believed and those which are valued, and that it permits in a large class of cases the assignment of both belief and value to the same object. The advantage is not complete, however, since no definition of belief is provided in the case of propositions which are of ultimate value. If $a$ is of such importance to one that there are no $x$, $y$, and $z$ satisfying the condition of the definition, then there is no way of applying Ramsey's methods to measure belief in it. If, for example, $A$ is the foundation of values for the subject, then all $z$'s will be indifferent in the absence of $A$. Thus people who are very religious may be unable to contemplate the question of the existence of God as a hypothesis figuring in a wager, because they are convinced that nothing would be worth anything were it false. Nor must we employ metaphysical propositions to get this effect; a person whose life is founded in that of his family may be unable to entertain the hypothesis that they should all save him be suddenly killed. Not only may he be unable to do this because of a profound dislike for considering horrible things, but he may realize that his preferences would be essentially different in such an event; and be quite unable to contemplate the value to be attached to outcomes which depend upon and accompany it.

It may be that this failure of the Ramseyian method is a virtue of it, and that in cases when preference is ultimately strong, belief cannot appropriately be considered. The symmetrical form of this view, that the objects of ultimate belief are not appropriate objects of preference of value, has some plausibility. If scientific laws are taken as objects of ultimate belief,[15] then their function is more like that of norms than like that of ordinary propositions cognitively considered. It may be that the distinction between knowledge and value cannot be clearly made when either is of a very fundamental sort.

Yet another difficulty of applicability of the Ramseyian method occurs in cases of propositions in which one has or may have a fairly clear degree of belief but which are not measurable by the method. A man may certainly build a house for his family and descendants from one material rather than another because he deems the first more likely to endure than the second, though he is clear that both will last well beyond his death. If we think of this as a gamble at all, then we must think of the payoff as being to persons other than the agent, and Ramseyian utility functions are not as they stand capable of dealing with such cases.

### C. DIFFICULTIES ABOUT THE PRESUPPOSITIONS OF THE THEORY OF PREFERENCE

The first of these difficulties is that the preference axioms are apparently not in general true. In particular, indifference may not always be transitive. This point may, I think, be argued in either direction. A deeper difficulty is that propositions may be not comparable in respect of preference or indifference; a man may not properly be said to have a preference, for example, between the proposition that his wife will leave him and the proposition that he will contact an unspecified incurable disease. He cannot be said to be indifferent between these, nor are they propositions not of serious import. It may be that for some ways in which they could come true he prefers one to the other (the disease in question is a mild allergy to a rare plant), and for other ways this preference is reversed, but that he has no preference between them as stated.

This difficulty becomes even more extreme when we consider complex gambles. A man's willingness to gamble need not, in order to accord with reasonableness, be closed under, for example, disjunctions and conjunctions. The arguments in the first section of this chapter are intended to show this.

Another difficulty in the presuppositions of the theory of preference is that in this theory no allowance is made for the ways in which an agent's preferences may depend essentially upon other factors. Ramseyian utility theory considers the preferences of the subject as so many data. Yet, in fact, more frequently than not, what a man's preferences are cannot be determined except in concert with the preferences of others. For people who live in a harmonious group, unanimity of preference among the members of the group is of great importance to them, and one's preference may be ambiguous among various distinct orderings when this factor is ignored. There are several ways in which the theory may try to meet this challenge. The first of these is to attempt to extend the preference orderings of individuals to be defined for propositions in which reference is made to preference orderings, both those of the individual in question and those of other agents. This seems not very promising in view of the obvious difficulties with impredicativity which seem also essential. In cases such as those in question unanimity is not just another proposition to be considered in the ordering, it is a part of the foundation for the formation of preference and desire.

A second way to try to cope with this problem is to say that the preferences change in discussion with one's comrades. This pretty obviously does not fit the facts, the truth is that one's preferences are not defined in abstraction from the social context,[16] and to suppose that they are is to substantivize them in a way which is incompatible with the behavioristic bases of the Ramseyian theory. If preference is a parameter of behavior, then were behavior drastically different, preference could not be the same. And certainly, were the social context to be different, so would be the behavior.

## D. OTHER FEATURES

There are two other features of the Ramseyian theory which, though not difficulties in themselves, may be the seeds of difficulties and should be mentioned.

The first of these is that on Ramsey's view[17] the objects of desire are propositions, not individual objects or substances. This runs counter to the grammatical evidence of ordinary language according to which verbs of desire may, and in some cases virtually must, take grammatically substantive objects. 'Want' for example, characteristically takes a name or description of an individual substance as an object; (It may also, of course, take infinitive phrases.) This linguistic evidence is corroborated by the phenomenology of attitudes of desire or need; when one wants something what is before the mind is the thing, not some proposition about it.

Further, at least one important comprehensive theory of action, Aristotle's, has it that the object of desire or appetite is always a particular. The reasons for this in Aristotle's theory are complex and deep: The objects of desire must be the same in order to account for appetitive action, and the actions of animals must have particulars as their final and efficient causes.

Neither of these in itself is an objection to taking the objects of desire to be propositions, but they are both fairly weighty considerations, and they generate in an obvious way questions that a propositional theory must answer. There is for example the question of finding the appropriate propositional object to correspond to a desired individual object, and the problem of distinguishing desires which we should ordinarily count as desires for a specific object (He wants a car, that car.) from those which we count as being for an unspecified object (He wants a car; any car will do).

It should be remarked that this last is a question which is not as yet resolved by any account of desire, and it does not thus mark a difficulty of propositional theories in comparison with other accounts.

The great advantage of propositional theories is that they make possible the application of ordinary logics to the objects of desire, and in this way the logics of belief and desire may be unified. Whether this advantage outweighs their difficulties awaits a fuller theoretical development.

The other feature which may lead to difficulties in the Ramseyian theory of preference is that the theory does not allow the distinction of preferences according to intensity. This is analogous to Peirce's argument[18] that an adequate account of belief requires in addition to assigning a degree of belief to a proposition, also an assignment to indicate how secure or well founded that belief is. The extent to which you believe that a particular randomly selected coin is fair is, let us say; one-half. If you experiment with the same coin, flipping it, measuring it, etc., and come on the basis of this experiment to a belief of degree one-half that the coin is fair, then this more secure, better founded belief should be distinguished from the former, says Peirce. But the distinction is not one of degree of belief, and thus there is another dimension in which beliefs must be evaluated. These two dimensions can be thought of in many important and representative cases as the mean and standard deviation of a normal distribution.

In the case of preference we are liable to think that this difficulty is a consequence of the non-uniqueness of degree of preference, that is to say, of the relatively weak invariance characteristics of utility functions. The problem, it might be claimed, is that we have no clear meaning for strength of preference, and that what is needed is not a second dimension for measuring preferences, but a strengthening of the dimension that we have. If, however, we conform analogously to Peirce's argument about belief, then this claim is not quite to the point. We can still distinguish the degree of preference, or amount to which one object is preferred to another (even though we cannot measure this in a Ramseyian theory) from how secure or well founded that preference is. Indeed, the same point can be made about the preferential concept for which in the Ramseyian theory degrees are well defined, namely the ratio of preference intervals, or of preference mid-points: I may be indifferent between $x$ and an even-chance gamble on $y$ and $z$, having had no extensive experience of any of $x$, $y$ or $z$. After considerable experience of them I may still be indifferent between $x$ and the even chance gamble on $y$ and $z$, but the latter attitude is much more securely founded than the first. This foundation is not indicated in the ration of the $y-x$ interval to that of the $z-x$ interval, since this ratio is the same in both cases.

As with the former question of the nature of the objects of desire this does not seem to be an objection to the Ramseyian theory as much as it is an argument that the theory is incomplete as it stands. Here as in other respects we should be gentle in our criticism of that theory since it and its relations and descendants are just about the only articulated accounts of preference, desire, and belief in existence.

Although it is not a theory or account of belief, the theory of deliberation of R. C. Jeffrey should at least be mentioned here. In Jeffrey's theory the value or *desirability* of a proposition is taken to be a weighted average of propositions which entail it, the weights being the appropriate beliefs. Thus if $B_1, \ldots, B_k$ are pairwise inconsistent and if $A$ is equivalent to their disjunction, then the desirabilities are given by a function *des* satisfying

$$\mathrm{des}(A) = \sum_{i=1}^{k} p(B_i) \cdot \mathrm{des}(B_i)$$

where $p$ gives degree of belief. It follows from this that desirability conforms to the constraint of additivity. Not the least advantage of Jeffrey's approach, which is intended to be an account of deliberation, is that it avoids the first of the difficulties mentioned above, which grows out of the requirement on the part of Ramsey's theory that beliefs about wagers assume such a crucial role in it. Jeffrey[19] discusses this in terms of beliefs about *causality*. He charges Ramsey's account with depending upon unwarranted and unanalyzed assumptions about the agent's causal beliefs, and compares his own theory favorably with it in this respect. In connection with the present discussion, however, it should be emphasized that the aims of the two accounts differ in at least one important respect, namely that Ramsey was trying to give an account of belief, while Jeffrey was trying to provide an account of deliberation, and was not concerned with the psychological question of what it means to say that a man has a given belief. Of course this difference of intent is no shortcoming of either account.

## NOTES

[1] Cf. *Hume*, pp. 127ff.

[2] Writing the probability of $Y$ conditional upon $X$ as $P_X(Y)$.

[3] *Hume*, pp. 161f, 166.

[4] *Bayes*.

[5] The account of measurement in this section derives from that of *Scott and Suppes* and *Suppes and Zinnes*.

[6] As is done in Jeffrey's theory. There each proposition which is indifferent to its negation is assigned zero value. The result, in fact, is not completely unnatural. See *Jeffrey*, Chapter 5.

[7] See *Suppes and Zinnes*, Theorem 13.

[8] In 'Truth and Probability', in *Ramsey*. The paper is not easy reading, and Jeffrey gives a lucid summary of the method in Chapter 3 of *Jeffrey*.

[9] The axiomatization given here is an inessential variant of that found in *Suppes and Zinnes*, definition 13.

[10] Ramsey's axioms, when interpreted as intended, required continuity as well as density. Suppes was the first to give a set of axioms with plausible interpretations in the rational numbers. See *Suppes*.

[11] See *Scott and Suppes* for proofs.

[12] In the Aristotelean account of practical reason and action.

[13] See Chapter 10 of *Jeffrey* for a discussion of this and other problems, some of which are overcome in Jeffrey's theory.

[14] Over, for example, that of Savage, which does presume such a division.

[15] Such an account is proposed by Kuhn.

[16] See the essay 'Hedonism' in *Marcuse*.

[17] And on Jeffrey's. See section 4.1 of *Jeffrey*.

[18] In *Peirce*.

[19] Pp. 145–150.

CHAPTER III

# LOGIC AND PROBABILITY

## III.1. LOGIC

We shall assume that the objects of belief are or correspond to sentences. These sentences should not be thought of as uninterpreted, they are assumed to have determinate meanings. We understand a sentence to be much like what logicians have traditionally referred to as a *proposition*; it is the bearer of truth or falsity, the intended object of thought, what is contemplated in an assertoric judgment, the content of an assertion, and so on.

Sentences are taken to be made up of *predicates* of varying degrees, of individual *names*, of truth-functional sentential connectives, and of quantified individual variables. If $R$ is an $n$-place predicate and $a_1, \ldots, a_z$ are individual names, then $R(a_1, \ldots, a_n)$ is an atomic sentence. If $v$ is an individual variable then $\exists vR(a_1, \ldots v \ldots, a_n)$ and $\forall vR(a_1, \ldots v \ldots, a_n)$ are *existential* and *universal quantifications* respectively of $R(a_1, \ldots, a_n)$. If $A_1$ and $A_2$ are sentences then $-A_1$, $A_1 \vee A_2$, and $A_1 \wedge A_2$ are respectively the *negation* of $A_1$, and the *disjunction* and *conjunction* of $A_1$ and $A_2$. $A_1 \rightarrow A_2$ is the *conditional* of which $A_1$ is the *antecedent* and $A_2$ the consequent, and $A_1 \leftrightarrow A_2$ is the *biconditional* of $A_1$ and $A_2$. These logical operations are understood in their usual senses, as are the standard concepts of propositional and predicate logic: Some sentences are tautologous, some tautologically inconsistent, some are predicatively or first-order valid and some first-order inconsistent, and the clarity of these and related notions is assumed. In particular we allow also that these logics may be extended. And in general by a *logic* $T$ is comprehended a related cluster of notions: $T$-implication, $T$-necessity, $T$-incompatibility, $T$-consistency, etc. These are taken to be related as follows. In each case $X$ and $Y$ are sets of sentences and $A$ is a sentence.[1]

$X$ *T-implies* $A \Leftrightarrow X \cup \{-A\}$ is $T$-inconsistent
$X$ *T-implies* $Y \Leftrightarrow X$ $T$-implies every member of $Y$
$A$ is *T-necessary* $\Leftrightarrow -A$ is $T$-inconsistent
$X$ is *T-necessary* $\Leftrightarrow$ Every member of $X$ is $T$-necessary.

In every case $T$ is assumed to be an extension of the ordinary truth-functional logic of propositions, in the sense that all tautologies are $T$-necessary, and all tautologically inconsistent sets are $T$-inconsistent, and also that $T$ is absolutely consistent, that is, that for no sentence $A$ are both $A$ and its negation $T$-necessary. In most cases $T$ may be assumed also to be an analogous extension of predicate logic, and where this is not so, or where it is required, explicit notice is given.

No distinction is made in terms of the axiomatizability characteristics of logics as such, although many such distinctions are entailed. The concept of *compactness* in its ordinary logical sense does, however, turn out to be of considerable importance for the relation of logic and probability, and in many places logics are distinguished which are *compact* for inconsistency: $T$-inconsistency is *compact* (in the set of sentences $\Omega$) if every $T$-inconsistent set of sentences (subset of $\Omega$) has a finite $T$-inconsistent subset.

In brief, no confusion should result if the logics $T$ are thought of as first-order theories, in the ordinary comprehension of that term, without prejudice as to the axiomatizability characteristics of those theories; although it should not be required that logical notions be understood in this way. In many cases the essentially predicative features of the logics $T$ are not exploited, and in view of this they are in general assumed to be only at least tautological and absolutely consistent.

Now let $\Omega$ be a denumerable collection of sentences which includes every finite truth-functional combination of any of its members ($\Omega$ is said to be a *field* of sentences). If $X$ is a finite subset of $\Omega$ then $\bigwedge X$ and $\bigvee X$ are respectively the conjunction and disjunction of the members of $X$. By a *constitution* of the (perhaps denumerably infinite) subset $X$ of $\Omega$, which subset may be $\Omega$ itself, is meant a set which includes for each sentence $A \in X$ either $A$ or its negation. The set $X$ is said to be *complete in* $\Omega$ *or maximal in* $\Omega$ if for every member of $\Omega$ either it or its negation is a member of $X$. Clearly the maximal subsets of $\Omega$ are just the constitutions of $\Omega$. An important characteristic of logics $T$ such as we are considering is given in *Lindenbaum's Theorem*:[2]

If $\Omega$ is a denumerable field of sentences, and $X$ is a $T$-consistent subset of $\Omega$, then $X$ is a subset of some maximal $T$-consistent subset of $\Omega$. (Every $T$-consistent subset of $\Omega$ has a $T$-consistent extension which is complete in $\Omega$.)

Given a field $\Omega$ of sentences, and the collection $M(\Omega)$ of maximal

$T$-consistent subsets of $\Omega$, the *Tarski–Lindenbaum Algebra* (The T–L Algebra) of subsets of $M(\Omega)$ is formed as follows: For each sentence $A \in \Omega$, let $f(A)$ be the set of all those maximal $T$-consistent subsets of $\Omega$ (members of $M(\Omega)$) each of which includes $A$ as a member. Then

(i) $f(-A) = M(\Omega) - f(A)$
(ii) $f(A \vee B) = f(A) \cup f(B)$
(iii) $f(A \wedge B) = f(A) \cap f(B)$
(iv) If $A$ is $T$-necessary then $f(A) = M(\Omega)$
(v) If $A$ is $T$-inconsistent then $f(A) = \Lambda$
(vi) If $A$ and $B$ are $T$-equivalent, then $f(A) = f(B)$
(vii) If $f(A) = f(B)$ then $A$ and $B$ are $T$-equivalent.
(viii) If $A$ $T$-implies $B$ then $f(A) \subseteq f(B)$
(ix) If $f(A) \subseteq f(B)$ then $A$ $T$-implies $B$.
(x) The set of images of members of $\Omega$ under $f$ (those subsets $f(A)$ of $M(\Omega)$ where $A \in \Omega$) is a Boolean algebra of sets.

The function $f$ described above which assigns

$$\{\Phi \mid \Phi \text{ is a maximal } T\text{-consistent extension of } \{A\}\}$$

to the sentence $A \in \Omega$ is said to be the *canonical map* of $\Omega$ on to the *T–L* Algebra $\mathfrak{B}(\Omega, T)$.

Given a field $\Omega$ of sentences, the $T$-consistent constitutions of $\Omega$, the maximally $T$-consistent subsets of $\Omega$, are; if assertable, the strongest or most determinate $T$-consistent assertions possible in terms of $\Omega$. They correspond to the atoms of a set-theoretical field. It is also important to characterize the largest classes of internally independent (and not $T$-necessary) sentences in terms of $\Omega$. A set of sentences is internally independent in this sense if no one of them and no negation of any of them follows from the rest: One might assert any sub-class of them and yet $T$-consistently assert or deny any other of them. Such sentences will correspond to the basic sets of set theoretical fields. To characterize them we describe first certain collections of them.

The set $X \subseteq \Omega$ is a *finite T-base* for the field $\Omega$ if $X$ is finite, and further

(i) Every constitution of $X$ is $T$-consistent.

(ii) If $A$ is any $T$-consistent sentence in $\Omega$ then there is a collection $C_A = \{X^1, \ldots, X^k\}$ of constitutions of $X$ such that $A$ is $T$-equivalent to $\wedge X^1 \vee \wedge X^2 \vee \ldots \vee \wedge X^k$.

And the field $\Omega$ is *essentially T-finite* if it has a finite $T$-base.

If $X$ is a finite $T$-base for $\Omega$ then the members of $X$ are *T-basic sentences*; Some consequences may help to clarify these concepts.

THEOREM 1.1. If $X$ is a finite $T$-base for $\Omega$, then $X$ is $T$-equivalent to any of its $T$-consistent supersets which are subsets of $\Omega$;

*Proof*: Let $Y$ be a $T$-consistent superset of $X$ which is a subset of $\Omega$. Then $Y$ $T$-implies $X$, and it must be shown that $X$ $T$-implies $Y$, that is, that $X$ $T$-implies each member of $Y$. Let $A \in Y$, and let $C_A = \{X^1, \ldots, X^\kappa\}$. Then $A$ is $T$ equivalent to $\wedge X^1 \vee \ldots \vee \wedge X^\kappa$. Since $X \cup \{A\} \subseteq Y$, and $Y$ is $T$-consistent, $X \cup \{A\}$ is $T$-consistent. Thus

$$\wedge X \wedge (\wedge X^1 \vee \ldots \vee \wedge X^\kappa)$$

is $T$-consistent. Thus for some $X^l$, $\wedge X \wedge \wedge X^l$ is $T$-consistent. Since both $X$ and $X^l$ are constitutions of $X$, if they differ then some sentence is a member of one of them, and its negation is a member of the other. But since $X \cup X^l$ is $T$-consistent, this cannot be so. Thus $X = X^l$ and $X \in C_A$. Thus $X$ $T$-implies $C_A$, and hence $T$-implies $A$.

THEOREM 1.2. If $X$ and $Y$ are both finite $T$-bases for $\Omega$ and $X \subseteq Y$ then $X = Y$.

*Proof*: Suppose $X$ and $Y$ to be distinct. By Thm. 1.1. they are $T$-equivalent, and, since they are distinct, there is some $A \in Y - X$. Since $Y$ is a $T$-base, $(Y - \{A\}) \cup \{-A\}$, being a constitution of $Y$, is also $T$-consistent, and is a superset of $X$. Hence, by Thm. 1.1. this set is $T$-equivalent to $X$. Thus $Y$ is $T$-equivalent to $(Y - \{A\}) \cup \{-A\}$, which violates the assumption that $Y$ is $T$-consistent.

THEOREM 1.3. If $X$ is a finite $T$-base for $\Omega$ and $A \in \Omega$, then either $X$ $T$-implies $A$ or $X$ $T$-implies $-A$.

*Proof*: If $X \cup \{A\}$ is $T$-inconsistent, then $X$ $T$-implies $-A$. If, on the other hand, $X \cup \{A\}$ is $T$-consistent, then by Thm. 1.1, $X$ is $T$-equivalent to it, and hence $X$ $T$-implies $A$.

THEOREM 1.4. If $X$ is a finite $T$-base for a denumerable field $\Omega$ then every maximal $T$-consistent subset of $\Omega$ is $T$-equivalent to a unique constitution of $X$ and conversely.

*Proof*: We argue first that to each maximal $T$-consistent subset of the field $\Omega$ there corresponds a unique $T$-equivalent constitution of the finite $T$-base $X$.

Let $Y$ be a maximal $T$-consistent subset of $\Omega$. Then $Y$ is a $T$-consistent constitution of $\Omega$. Hence for every $A \in X$, either $A$ or $-A$ is a member of $Y$. Thus some constitution of $X$ is a subset of $Y$. Since $Y$ is $T$-consistent, this constitution is unique, for distinct constitutions are $T$-inconsistent. By Thm. 1.1 $X$ is $T$-equivalent to $Y$.

Now to see the correlation in the opposite direction; let $X^l$ be a constitution of the $T$-base $X$. $X^l$ is $T$-consistent and hence the collection $Y \subseteq \Omega$ of propositions $T$-implied by $X^l$ is $T$-consistent. We can see that this collection is a constitution of $\Omega$, for let $A \in \Omega$. Then by Thm. 1.3 either $X$ $T$-implies $A$ or $X$ $T$-implies $-A$. Thus either $A \in Y$ or $-A \in Y$. Since $Y$ is $T$-consistent, not both $A \in Y$ and $-A \in Y$, and hence $Y$ is a $T$-consistent constitution of $\Omega$. Thus $Y$ is a maximal $T$-consistent subset of $\Omega$ and is unique, for the class of propositions $T$-implied by $X$ is unique.

COROLLARY. If $X$ is a finite $T$-base for $\Omega$ and $X^l$ a constitution of $X$, then there is some maximal $T$-consistent subset of $\Omega$, $\Phi_j$, such that

$$\Phi_j = \{A \mid X_j \ T\text{-implies } A\}.$$

*Proof*: Let $\Phi_j$ be the maximal $T$-consistent subset of $\Omega$ to which $X$ is $T$-equivalent. Every $A \in \Omega$ which is a $T$-consequence of $\Phi_j$ is a member of $\Phi_j$, and $X^l$ and $\Phi_j$ have the same members of $\Omega$ as $T$-consequences.

THEOREM 1.5. (Cardinality theorem.) If $\Omega$ is a denumerable field with a finite $T$-base $X$ of size $k$, then there are exactly $2^k$ $T$-consistent constitutions of $\Omega$.

*Proof*: Directly from Thm. 1.3 by the number of subsets of a set of given finite size.

The essentially finite nature of fields with finite $T$-bases is revealed also in the finiteness of the associated Tarski-Lindenbaum Algebra.

THEOREM 1.6. If the denumerable field $\Omega$ is essentially $T$-finite, then $\mathfrak{B}(\Omega,T)$ the associated $T$–$L$ Algebra, is a finite Boolean Algebra. If $\Omega$ has a finite $T$-base of size $k$, then $\mathfrak{B}(\Omega,T)$ has just $2^{2^k}$ elements.

*Proof*: If $X$ is a $T$-base for $\Omega$ of size $k$, then $X$ has just $2^k$ $T$-consistent constitutions, and hence, by Thm. 1.4 there are just $2^k$ distinct maximal $T$-consistent subsets of $\Omega$, call them $\Phi_1, \ldots, \Phi_2\kappa$. Now let $\Phi'_1, \ldots, \Phi'_n$ be any $n$ distinct $\Phi_l$. Each of these is $T$-equivalent to some constitution of the base $X$, so let $\Phi'_1$ be $T$-equivalent to $X_1, \ldots, \Phi'_n$ be $T$-equivalent to $X_n$.

Then the sentence

$$A = \wedge X_1 \vee \ldots \vee \wedge X_n$$

is $T$-implied by each of $\Phi'_1 \ldots, \Phi'_n$ and is hence a member of each of $\Phi'_1, \ldots, \Phi'_n$. Further if $\Phi_l$ is any maximal $T$-consistent subset of $\Omega$ which is distinct from all of $\Phi'_1, \ldots, \Phi'_n$, then $\Phi_l$ is $T$-inconsistent with each of them, hence with each of $X_1, \ldots, X_n$, and hence with the sentence $A$. Thus this sentence, $A$, is a member of no other $\Phi_l$, and

$$f(A) = \{\Phi'_1, \ldots, \Phi'_n\}.$$

This shows that every non-null subset of the collection $\{\Phi_1, \ldots, \Phi_{2\kappa}\}$ is the image of some member of $\Omega$ by $f$. Of course the null set is the image of each $T$-inconsistent member of the algebra $\mathfrak{B}(\Omega, T)$, and hence this algebra has just $2^{2^k}$ members.

THEOREM 1.7. If $\Omega$ is a denumerable field of sentences, and the subsets $X$ and $Y$ are subsets of just the same maximal $T$-consistent subsets of $\Omega$ then $X$ and $Y$ are $T$ equivalent.

*Proof*: Assume $X$ and $Y$ to be not $T$-equivalent. We may assume without loss of generality that some member of $Y$ is not $T$-implied by $X$. That is, that for some $A \in Y$, $X \cup \{-A\}$ is $T$-consistent, and hence has a maximal $T$-consistent extension. This maximal subset of $\Omega$ is not a superset of $Y$.

If $\Omega$ is a field with a finite $T$-base $X$, and if $Y$ is also a $T$-base of $\Omega$, then $X$ and $Y$ are of the same cardinality, and there is a one-to-one correspondence between the constitutions of $X$ and those of $Y$, in which corresponding constitutions are $T$-equivalent. This is a direct consequence of Theorem 1.4.

## III.2. The probability of sentences

From a measure theoretic point of view, the study of probability is the study of a certain clearly defined class of measures. From the point of view of epistemology, and the theory of judgment generally, the study of probability is the study of the logics of judgments of varying strength. A question which is central to our general understanding of probability is this: *How are judgments of varying strength structured so that the measure theoretic concept of probability applies to them?* This question is important, it may be said to define the problems of the foundations of probability, both from the side of measure theory and from that of the logic of judgments. It is important to measure theory, because the relation to partial judgment is the defining character of probability theory, and in the absence of such relation probability measures would be no more than a class of mathematical functions of certain characteristics, of no special interest outside of mathematics, and not to be distinguished in an essential way from other classes of functions. This essential relation to judgment obtains as well when the laws of probability are interpreted as truths about frequencies or propensities. If we ask why there should be a mathematical theory of probability, the answer must be that there are important structural questions about the logical relations among partial judgments, and that since the structure of such judgments is described by probability measures, it is to the theory of those measures that we should look for answers to those questions.

The question of the relation of measure theoretic probability to the study of judgment is important to the study of judgment because the end of that study is a general theory of judgment, and it is only through description and investigation of the structure of judgments that sufficient generality can be achieved. That is, in general, how the logic of apodictic, or non-partial, judgment has proceeded, and it is only through the revelation of structure that the logic of deductive relations has advanced.

A paradigm answer to the question of the relation of probability to judgment is given by Hume in the passage from the *Treatise* discussed in the preceding chapter. This answer has its difficulties, but it cannot be faulted in its direction: It is an obvious and explicit attempt to show how partial judgment has the structure of a probabilistic object.

The response to the question given in this book is in general that of contemporary authors.[3] It is, first, that the objects of judgment are, or are expressed by, sentences, and that the logical relations among judgments are mirrored by those among the sentences which express them. Given a field of sentences and a logic which structures them, the Tarski-Lindenbaum Algebra classes of maximal *T*-consistent sets is a Boolean Algebra. This algebra is a domain for probability measures in the measure theoretic sense, and a probability on a field of sentences is to be understood in terms of a probability on the associated Tarski-Lindenbaum Algebra. That is the general approach, some ramifications of it may be worth remarking:

(1) Different logics, extensions of classical propositional logic, give different Tarski-Lindenbaum Algebras for the same field of sentences. These give different probabilities, different classes of probability measures

on the field of sentences. The concept of probability will be relativized to a logic $T$.[4] A part of the philosophical import of this is that we can approach the question of the reasonableness of partial belief as a question about the nature of the base logic $T$, and not, as certain authors, notably Carnap,[5] have approached it, as the question of what constraints in addition to the laws of probability should be put upon reasonable belief. All partial belief, reasonable or not, obeys some logic, in the sense that even to describe it requires a probability measure on a Tarski-Lindenbaum Algebra.[6] The standards for reasonable belief are not laws in addition to those of probability, they are achieved by strengthening the base logic.

(2) From a measure theoretic point of view, probability measures are defined not on Boolean Algebras in general, but on sigma-algebras, algebras closed under denumerably infinite as well as finite unions. This raises the epistemological question of the proper understanding of judgments with infinitely complex objects. We turn to this question in a later section.

(3) An important question in partial judgment is that of transparency or invariance of strength of judgment. Can we, for example, distinguish partial beliefs in logically equivalent sentences? The next section involves the initial consideration of this question.

Given a field $\Omega$ of sentences, an absolutely consistent and at least tautological logic $T$, and the associated $T$-$L$ Algebra $\mathfrak{B}(\Omega,T)$ of classes, we can define a $T$-probability measure on $\Omega$ in either of two equivalent ways. The first is to take as fundamental probability on the $T$-$L$ Algebra $\mathfrak{B}(\Omega,T)$. Then, given $\Omega,T$, and the representing function $f$, a *finite probability* on $\mathfrak{B}(\Omega,T)$ is a function $\mu$ such that

(1) $0 \leqslant \mu(\alpha) \leqslant 1$ for every $\alpha \in \mathfrak{B}(\Omega, T)$

(2) $\mu(M(\Omega)) = 1$

(3) If $\alpha \cap \beta = \Lambda$ then $\mu(\alpha \cup \beta) = \mu(\alpha) + \mu(\beta)$.

And a probability $p$, which assigns values to the sentences in $\Omega$ may be defined in terms of $f$ and $\mu$:

DEFINITION: $p$ is a *finite probability* on $\Omega$ if and only if for some finite probability $\mu$ on $\mathfrak{B}(\Omega,T)$, for all $A \in \Omega$

$$p(A) = \mu(f(A)).$$

The second method is to take the probability of sentences in $\Omega$ as fundamental and to characterize this without explicit dependence upon the algebra $\mathfrak{B}(\Omega,T)$.

SECOND DEFINITION: If $\Omega$ is a denumerable field of sentences, and $T$ is an absolutely consistent and at least tautological logic, then a finite *T-probability* on $\Omega$ is a function $p$ such that

(1) $0 \leqslant p(A) \leqslant 1$ for every $A \in \Omega$.
(2) If $A$ is $T$-necessary, then $p(A) = 1$.
(3) If $A$ and $B$ are $T$-incompatible, then $p(A \vee B) = p(A) + p(B)$.

The following theorem establishes the equivalence of these definitions.

THEOREM 2.1. If $\Omega$ is a denumerable field of sentences, $T$ an absolutely consistent and at least tautological logic, $\mathfrak{B}(\Omega,T)$ the $T$-$L$ Algebra associated with $\Omega$ and $T$, and $p$ is a function on $\Omega$, then the following two sets of conditions are equivalent.

(1.1) $0 \leqslant p(A) \leqslant 1$ for every $A \in \Omega$.
(1.2) If $A \in \Omega$ is $T$-necessary, then $p(A) = 1$.
(1.3) If $A$ and $B$ are $T$-incompatible members of $\Omega$ then $p(A \vee B) = p(A) + p(B)$.
(2) There is a finite probability $\mu$ on the algebra $\mathfrak{B}(\Omega,T)$ and for each $A \in \Omega$, $p(A) = \mu(f(A))$ where $f$ is the canonical map of $\Omega$ to $\mathfrak{B}(\Omega,T)$.

*Proof*: First assume the conditions (1). Then if $A$ and $B$ are $T$-equivalent, both are $T$-equivalent to $A \wedge B$, and both $A \wedge -B$ and $-A \wedge B$ are $T$-inconsistent. Since $T$ is at least tautological,

$$(A \wedge B) \vee (A \wedge -B) \vee (-A \wedge B) \vee (-A \wedge -B)$$

is $T$-necessary, and hence

$$(A \wedge B) \vee (-A \wedge -B)$$

is $T$-necessary. Thus both

$$A \vee (-A \wedge -B)$$
$$B \vee (-A \wedge -B)$$

are $T$-necessary, and

$$p(A \vee (-A \wedge -B)) = 1 = p(B \vee (-A \wedge -B)).$$

The members of each disjunction are tautologically; hence $T$-incompatible, so

$$p(A) = 1 - p(A \wedge -B) = p(B).$$

Thus $p$ is invariant for sentences in the same $T$-equivalence class. Further, if $f(A) = f(B)$ then $A$ and $B$ are $T$-equivalent, thus if $f(A) = f(B)$ then $p(A) = p(B)$, and we can consistently define a function $\mu$ on $\mathfrak{B}(\Omega, T)$:

$$\text{If } \alpha \in \mathfrak{B}(\Omega, T), \text{ let } \mu(\alpha) = p(A)$$
$$\text{where } A \in \Omega \text{ and } \alpha = f(A).$$

Now it remains to show that $\mu$ is a probability on $\mathfrak{B}(\Omega, T)$.

Clearly $\mu$ is everywhere defined on $\mathfrak{B}(\Omega, T)$, for each member of $\mathfrak{B}(\Omega, T)$ is the image of some $A \in \Omega$ by the canonical map, and we assume $p$ to be everywhere defined on $\Omega$. Similarly we have that

$$0 \leqslant \mu(\alpha) \leqslant 1 \text{ for } \alpha \in \mathfrak{B}(\Omega, T).$$

Since the $T$-necessary members of $\Omega$ are mapped by $f$ to $M(\Omega)$, and $p$ assigns 1 to each such sentence

$$\mu(M(\Omega)) = 1.$$

To assure that $\mu$ is finitely additive, let $\alpha$ and $\beta$ be disjoint members of $\mathfrak{B}(\Omega, T)$, and let $f(A) = \alpha$ and $f(B) = \beta$. Then

$$f(A \vee B) = \alpha \cup \beta.$$

Since $\alpha$ and $\beta$ are disjoint, no maximal $T$-consistent extension of $A$ is also a maximal $T$-consistent extension of $B$. Thus $\{A, B\}$ has no maximal $T$-consistent extension, and; by Lindenbaum's theorem, is not $T$-consistent. Thus

i.e.; $$p(A \vee B) = p(A) + p(B)$$
$$\mu(\alpha \cup \beta) = \mu(\alpha) + \mu(\beta)$$

which establishes that the conditions (1) are sufficient for (2).

Now to establish the converse, assume (2). Then, since the canonical map is everywhere defined on $\Omega$ and $\mu$ is a finite probability, it is evident that $p$ satisfies (1.1).

If $A$ is $T$-necessary then $f(A) = M(\Omega)$ and $\mu(f(A)) = 1$ so $p(A) = 1$ and $p$ satisfies (1.2).

Finally, if $A$ and $B$ are $T$-incompatible, they share no maximal $T$-consistent extensions, so $f(A) \cap f(B) = \Lambda$ and we have $p(A \vee B) =$

$\mu(f(A \vee B)) = \mu(f(A) \cup f(B)) = \mu(f(A) + (f(B)) = p(A) + p(B)$. So $p$ satisfies (1.3) and the equivalence of the two definitions is established.

This theorem, and its generalization to infinite form which is considered later, is a fundamental truth of the relations of epistemology and the mathematics of probability. At the least, it provides a precise answer to the question how the measure theoretic laws of probability apply to sentences. Some of the fruits of the correlation which is revealed in the theorem may be seen by relating the *invariance* of probability measures to the epistemic concept of *transparency*.

## III.3. Transparency

Russell and Whitehead, in an appendix to the second edition of *Principia Mathematica*,[7] introduced the notion of transparency to distinguish those occasions on which a proposition is used to make an assertion about facts which themselves do not at all involve the proposition, from those on which a proposition is used to make an assertion about facts or objects among which may be the proposition itself. In the former case the instance of the proposition is said to be *transparent*, one sees through it the facts or objects about which the assertion is made; while in the latter case the instance is said to fail of transparency; what one sees is in part the proposition itself.

If an instance of a proposition is transparent, then the truth-value of the assertion in which it occurs is unaffected when that proposition is replaced by any equivalent proposition. Should such replacement change the truth-value of the assertion, then that assertion must be in part an assertion about the proposition in question. Russell and Whitehead cite the occurrence of propositions in truth-functions as examples of transparent occurrence – the truth-value of a truth-function of propositions is unaffected when any of the involved propositions is replaced by one of the same truth-value – and they give the occurrence of a proposition '$p$' in '$A$ believes $p$' as an example of non-transparent occurrence; $A$ may believe $p$ without believing equivalent propositions.

The same proposition may be transparent on one occurrence and not so in other instances, and it is thus in features of the instance, and not in the proposition itself, that the root of transparency, or its failure *opacity*[8] must reside. Thus Quine and other writers have come to speak of *contexts* or *modes of containment* which form sentences with designative expressions as opaque or transparent: If $S$ is a context or mode of containment which forms a sentence $S(t)$ whenever $t$ is a singular term, then $S$ is *opaque* if there are terms $t_1$ and $t_2$ such that $t_1 = t_2$ but such that the sentences $S(t_1)$ and $S(t_2)$ differ in truth-value. This definition is applied to predicates and sentences using coextensionality and sameness of truth-value respectively in the place of identity. Contexts which are not opaque are said to be *transparent*.[9]

This definition may be generalized by allowing consideration of other equivalence relations: Thus if $T$ is any equivalence relation of sentences the function $f$ of sentences may be said to be *T-transparent* if $f$ is invariant under replacement of *T-equivalents*. That is simple enough and it allows us to turn a classificatory concept – transparency – into a comparative concept, the function $f$ being *at least as transparent as* the function $g$ if $f$ is transparent for every $T$ for which $g$ is transparent.

Our concern here will be restricted to functions of sentences which are probability measures and to equivalence relations which are logical equivalences in the sense characterized in section 1 of this chapter. If $p$ is a probability measure and $T$ is a logic, then $p$ is said to be $T$-transparent if $p$ is invariant under the replacement of $T$-equivalents.

One of the epistemically interesting consequences of theorem 2.1 and the association of measure-theoretic and epistemic probability which it reveals, is that any measure which is at least a tautological probability measure and is also $T$-transparent, is as a consequence also a $T$-probability. This is from the mathematical point of view a fairly obvious and simple point, and it is worth a brief development.

Let $\Omega$ be a denumerable field of sentences, and $T_0$ an absolutely consistent and at least tautological logic. Let $T_1$ be an *extension* of $T_0$, that is to say, whatever is $T_0$ necessary or $T_0$-inconsistent is also, respectively, $T_1$-necessary or $T_1$-inconsistent. Thus, every maximal $T_1$-consistent subset of $\Omega$ is also a maximal $T_0$-consistent subset of $\Omega$. $M_1(\Omega) \subseteq M_0(\Omega)$.

Now consider the algebras $\mathcal{B}(\Omega, T_0)$ and $\beta(\Omega, T_1)$, and the canonical maps $f_0$ and $f_1$, which map $\Omega$ onto them. Since $T_1$ is an absolutely consistent extension of $T_0$, if $A \in \Omega$; then

$$\begin{aligned} f_1(A) &= \{\Phi \mid A \in \Phi \in M_1(\Omega)\} \\ &= \{\Phi \mid A \in \Phi \in M_1(\Omega)\} \cap M_0(\Omega) \\ &= \{\Phi \mid A \in \Phi \in M_0(\Omega)\} \cap M_1(\Omega) \\ &= f_0(A) \cap M_1(\Omega) \end{aligned}$$

and in general the algebra $\mathcal{B}(\Omega, T_1)$ is just the result of intersecting the members of $\mathfrak{B}(\Omega,T_0)$ each with $M_1(\Omega)$. This is sometimes expressed by saying that $\mathfrak{B}(\Omega,T_1)$ is the *quotient algebra* of $\mathfrak{B}(\Omega,T_0)$ with $M_1(\Omega)$, written

$$\mathfrak{B}(\Omega,T_1) = \mathfrak{B}(\Omega,T_0)/M_1(\Omega).$$

Suppose now that $p$ is a $T_0$ measure on $\Omega$. Then $p$ associates with a probability $\mu_0$ on $\mathfrak{B}(\Omega,T_0)$

$$p(A) = \mu_0(f_0(A)), \quad \text{for every } A \in \Omega.$$

If $p$ is also $T_1$ transparent, that is, if $p(A)=p(B)$ whenever $A$ and $B$ are $T_1$-equivalent members of $\Omega$, then the probability $\mu_0$ will also have an invariance characteristic namely:

If $\alpha$ and $\beta$ are members of $\mathfrak{B}(\Omega,T_0)$ such that

$$\alpha \cap M_1(\Omega) = \beta \cap M_1(\Omega), \quad \text{then} \quad \mu_0(\alpha) = \mu_0(\beta).$$

*Proof* of this last: If $\mu_0(\alpha)\neq\mu_0(\beta)$; then if $A$ and $B$ are any members of $\Omega$ with $f_0(A)=\alpha$ and $f_0(B)=\beta$, then $p(A)\neq p(B)$. And – since $p$ is $T_1$-transparent – we have that if $\mu_0(\alpha)\neq\mu_0(\beta)$ and $f_0(A)=\alpha$, $f_0(B)=\beta$, then $A$ and $B$ are not $T_1$-equivalent. Thus there is some maximal $T_1$-consistent $\Phi \in M_1(\Omega)$ which either includes $A \wedge -B$ or $B \wedge -A$. Hence either $\Phi \in f_0(A)-f_0(B)$ or $\Phi \in f_0(B)-f_0(A)$, and in either case $\alpha\cap M_1(\Omega)\neq \beta\cap M_1(\Omega)$.

This invariance characteristic of $\mu_0$ lets us argue that the function $\mu_1$ defined on the algebra $\mathfrak{B}_1(\Omega,T_1)=\mathfrak{B}(\Omega,T_0)/M_1(\Omega)$

$$\mu_1(\alpha \cap M_1(\Omega)) = \mu_0(\alpha)$$

is a probability on $\mathfrak{B}(\Omega,T_1)$.

(i) Since $\mathfrak{B}(\Omega,T_1)$ is the quotient algebra and $\mu_0$ is a probability on $\mathfrak{B}(\Omega,T_0)$, $\mu_1$ is everywhere defined with non-negative values no less than 1 on $\mathfrak{B}$ $(\Omega,T_1)$.

(ii) $$\mu_1(M_1(\Omega)) = \mu_1(M_0(\Omega) \cap M_1(\Omega)) = \mu_0(M_0(\Omega)) = 1.$$

Now let $\alpha'$ and $\beta'$ be disjoint members of $\mathfrak{B}(\Omega,T_1)$. Then there are $\alpha$ and $\beta \in \mathfrak{B}(\Omega.T_0)$ such that $\alpha'=\alpha\cap M_1(\Omega)$ and $\beta'=\beta\cap M_1(\Omega)$, and

$$\begin{aligned}\mu_1(\alpha' \cup \beta') &= \mu_1((\alpha \cup \beta) \cap M_1(\Omega)) = \mu_0(\alpha \cup \beta)\\ &= \mu_0(\alpha) + \mu_0(\beta) - \mu_0(\alpha \cup \beta)\\ &= \mu_1(\alpha') + \mu_1(\beta') - \mu_1(\alpha \cap \beta \cap M_1(\Omega))\\ &= \mu_1(\alpha') + \mu_1(\beta') - \mu_1(\wedge \cap M_1(\Omega))\\ &= \mu_1(\alpha') + \mu_1(\beta') - \mu_0(\wedge)\end{aligned}$$

and since $\mu_0$ is a probability on $\beta(\Omega,T_0)$, $\mu_0(\wedge)=0$ so

(iii) If $\alpha'$ and $\beta'$ are disjoint members of $\beta(\Omega,T_1)$ then $\mu_1(\alpha'\cap\beta')=\mu_1(\alpha)+\mu_1(\beta)$.

So $\mu_1$ is a probability on $\beta(\Omega,T_1)$. Further, the function $p$ on $\Omega$ will be related to $\mu_1$ in terms of the canonical map $f_1$: Since

$$p(A) = \mu_0(f_0(A))$$

and

$$f_1(A) = f_0(A) \cap M_1(\Omega)$$

we have that

$$p(A) = \mu_0(f_0(A)) = \mu_1(f_0(A) \cap M_1(\Omega)) = \mu_1(f(A))$$

which shows that $p$ is also a $T_1$-probability on $\Omega$.

This discussion may be summed up in a theorem relating transparency and probability:

THEOREM 3.1. Let $\Omega$ be a denumerable field of sentences, $T_0$ an absolutely consistent and at least tautological logic, and $T_1$ be an absolutely consistent extension of $T_0$. Then if $p$ is a $T_0$-probability on $\Omega$ which is also $T_1$-transparent, $p$ is also a $T_1$-probability on $\Omega$.

A useful and interesting consequence of this theorem is easy to come by;

THEOREM 3.2. If $\Omega$ is a denumerable field of sentences with a finite $T$-base $X$, and $p$ is a finite $T$-probability on $\Omega$, then for each $A \in \Omega$

$$p(A) = \sum_{X_l \in C_A} p(\wedge X_l)$$

where $C_A$ is the class of constitutions of $X$ each of which $T$-implies $A$;

*Proof*: Let $C_A=\{X_1,\ldots, X_\kappa\}$. Then $A$ is $T$-equivalent to

$$\wedge X_1 \vee \ldots \vee \wedge X_\kappa$$

$X_1,\ldots, X_\kappa$ are distinct, and hence $T$-incompatible, constitutions of $X$. By the previous theorem $p$ is $T$-transparent. Since it is also a $T$-probability, it is additive over finite disjunctions of pairwise $T$-incompatible sentences. Thus the theorem follows.

Transparency gives one way of understanding reasonableness of partial belief: A man's beliefs are unreasonable if he does not believe the consequences of what he believes. In the previous discussion this principle is ramified in two ways. First, 'believing the consequences of what he believes' is interpreted for partial belief as invariance of the extent of belief under logically equivalent transformations, or, what amounts to the same thing in at least tautological logics, as weak monotonicity of the extent of belief under logical implication; belief in a proposition may be no weaker than belief in its consequences. The second ramification of the principle is the relativization of the notion of logical equivalence or logical consequence to allow that these notions are not unique. Any concept of partial belief must be probabilistic in some sense, or the comparison of strength can have no meaning. It follows from this, as revealed in the preceding theorems, that every concept of partial belief must be transparent for some logic, in particular for the logic associated with the probability which gives its strength or degree. To attribute reasonableness or unreasonableness to the partial beliefs of an agent is to speak in terms of a logic which is stronger than, in the sense of being a proper extension of, the logic which is used to identify those beliefs probabilistically. It is in this way that we propose an answer to the question of whether probability, considered as the logic of partial belief, is descriptive, that is to say, that the laws of probability are *a priori* laws of belief, or, on the other hand, whether those laws are normative, in which case they constitute a set of ideals or canons of reasonableness which partial beliefs approach insofar as they are reasonable. Hume's account of partial belief is of the first of these two sorts. He is careful to build into his theory of the description of partial belief not only the laws of additivity over finite disjunctions of incompatible alternatives, and of unit probability for necessary propositions, but also to argue that the principle of indifference is an *a priori* character of belief, that no consistent description is possible of belief which violates it.

> Since therefore an entire indifference is essential to chance, no one chance can possibly be superior to another, otherwise than as it is compos'd of a superior number of equal chances. For if we affirm that one chance can, after any other manner, be superior to another, we must at the same time affirm, that there is something, which gives it the superiority, and determines the event rather to that side than the other: That is, in other words, we must allow of a cause, and destroy the supposition of chance; which we had before establish'd. A perfect and total indifference is essential to chance, and one total indifference can never in itself be either superior or inferior to another. This truth is not peculiar to my system, but is acknowledg'd by every one, that forms calculations concerning chances.[10]

More recent accounts of partial belief in terms of propensities to bet are of the other sort: They treat the laws of probability as norms or standards to which partial beliefs may, and should, approximate. Ramsey's theory may be cited as a paradigm of such views. Unreasonableness means, roughly, willingness to place bets on which, should they all be taken, one must be a net loser. (Or, in some accounts, on which there is no possibility of winning.)[11] These theories, and Ramsey's is perhaps more subtle in this respect than most, all, upon close examination, require some account of the description of partial beliefs. The problem is, in general, how to provide a consistent description of partial belief which is not *a fortiori* probabilistic and thus, by these accounts, already reasonable? This problem needs to be met not only for the salvation of these theories, but in a general setting. That is, one needs to be able to say what the difference is between belief and reasonable belief in a way which allows consistent descriptions of belief which is *not* reasonable. Hume's theory concentrates on the description of partial belief at the expense of reasonableness; if the canons of reasonableness are the laws of probability, then one cannot consistently describe within Hume's theory unreasonable partial beliefs. Ramsey's theory and others like it concentrate on the canons of reasonableness and leave the difficulties to the descriptive task. In both sorts of theories the distinction between belief and reasonable belief is undermined, or at least clouded.

The present account is not in opposition to these views, but is an attempt at a ramification of them. We may allow that all belief is probabilistic, and hence that the laws of probability are *a priori* or theoretical truths of the descriptive theory of belief. We may also hold that the laws of probability are the proper and complete canons of reasonableness for partial belief, and deny that these laws need supplementation by other conditions in order to provide an adequate normative logic of belief. This is made possible by distinguishing probabilities in terms of their logics, which may be done, as the preceding theorems show, either in terms of the logic upon which probability is based, or, equivalently, in terms of the transparency characteristics of partial belief. Then belief may be probabilistic in one sense; which is to say in terms of one logic, and not in another.

In the next chapter we commence with an abstract description of behavioristic theories of partial belief which relativizes them in analogous ways and relates them to the discussion of the present chapter.

## NOTES

[1] Tarski's paper 'Fundamental Concepts of the Methodology of the Deductive Sciences', in *Tarski* (1956), is the standard source for the development of these concepts. The expositions in *Van Frassen* and in *Schoenfield* may be easier for some to follow. The present chapter is intended to be self-contained, if a bit cryptic in places, with the remarked exception of some difficult and well-known proofs.

[2] See, for example, *Schoenfield*, p. 47.
[3] By Carnap in *Logical Foundations*, and in ramifications of Carnap's views in *Gaifman* and in *Scott and Krauss*. The present development is quite close to that of Scott and Krauss, and is indebted to their paper at several points.
[4] As far as I know, the first development of such a relativization was in my 1968 paper 'Probability and Non-Standard Logics', Carnap discusses the question, from a slightly different point of view, in section 5 of 'A Basic System of Inductive Logic' in *Carnap and Jeffrey*.
[5] In *Logical Foundations*, and in 'The Aim of Inductive Logic'.
[6] Compare this view with Quine's discussion of pre-logicality in *Quine* (1960).
[7] Volume I, Appendix C.
[8] This is Quine's term. It is used in 'Reference and Modality' and in *Quine* (1960) as well as elsewhere.
[9] Carnap in *Carnap*, (1956), uses *extensional* roughly as *transparent* is used here.
[10] *Hume*, p. 125.
[11] The stronger concept is developed in *Shimony*.

CHAPTER IV

# COHERENCE AND THE SUM CONDITION

## IV.1 The concept of coherence

In Chapter II the attempt to account for belief in terms of willingness to gamble was discussed. That discussion was inconclusive. It seems clear that belief is not reducible to such willingness or disposition, and it also seems clear that in at least some cases willingness to gamble in stakes of utility is at least an important consequence of believing.

In the present chapter betting functions and coherence are discussed in an informal and abstract way. The end is to relativize coherence to a notion of necessity, and to give a general structural characterization of it.

Suppose that we have a belief function defined in terms of utility intervals as described in II.2. Think for the moment of the utility of stakes of bets as being absolutely measurable, so that it makes sense to speak of a bet on a proposition with stakes of utility. If belief is associated with willingness to bet, then the belief function gives for each proposition $A$ for which it is defined the odds at which the believer will bet on $A$. If a man believes $A$ to degree $\frac{2}{3}$ then he will put up two units on condition that he receive three units if $A$ happens and nothing if $A$ does not happen, and he will accept two units on condition that he pay three units if $A$ happens and pay nothing if $A$ does not happen.

Here is one way in which such a believer may be unreasonable: He might believe $A$ to degree $\frac{2}{3}$ and also believe not-$A$ to degree $\frac{2}{3}$. At least a part of the force of calling this unreasonable is that – assuming the correlation of belief and willingness to bet – were the believer to bet on both $A$ and not-$A$, he would necessarily come out a loser, for he would put up $2+2=4$ units and would receive in any event just three units, this outcome depending not at all on which of the propositions $A$, not-$A$ occurred.

The essence of the concept of coherence is that a bettor must have some chance of not losing. That is to say that he must have some chance of recouping at least what he has put up. Perhaps the simplest case is that in which the stake of every bet is one unit. In this case if the bettor believes a proposition $A$ to degree $p(A)$ he will put up $p(A)$ on condition that he receive one unit if $A$ is true and zero if it is false. This is a consequence of the interpretation of belief in terms of a betting function, assuming a utility function. Such a function will assign the quantity

$$p(A) = \frac{x - z}{y - z}$$

to $A$, where $x$ is the value of the bet which yields $y$ if $A$ and $z$ if not-$A$. In the simple $1-0$ case, $y=1$ and $z=0$ so

$$p(A) = x$$

and the degree of belief in a proposition is just the value of the $1-0$ bet on that proposition.

Consider a collection of $1-0$ bets on the propositions in a finite set $X$ of size $n$. Suppose a bettor whose beliefs are defined on the members of $X$ bets on every proposition in $X$. Such a bettor puts up

$$\sum_{A \in X} p(A)$$

and should all and only the propositions in the subset $Y$ of $X$ obtain, he will receive one unit for each member of $Y$. That is to say, he will receive $N(A)$ units, where $N$ gives the size of its argument. If the bettor's beliefs as described by the function $p$ are to be coherent then it must be possible for him to receive at least the amount that he put up. In order for him to receive this amount there must be some subset $Y$ of $X$ such that

(i) $Y$ is conjunctively consistent.
(ii) $N(Y) \geqslant \sum_{A \in X} p(A)$.

That is to say:

(iii) A necessary condition for a function $p$ on a finite set $X$ of sentences to be coherent is that there be some consistent subset of $X$, the number of elements of which is not exceeded by the sum of the values of $p$ over $X$.

(iii) gives a necessary condition for coherence. It is not sufficient for several reasons: First the condition might hold for a set $X$, but not for all subsets of $X$. In this instance the bettor's beliefs might be unreasonable in several different respects, but so long as one considered only all the beliefs together, this unreasonableness might remain unmanifested. Here is an example.

$$\begin{array}{lcccccc} X = \{A \vee B, & A, & B, & -A, & -B, & -(A \vee B)\} \\ p: \quad \frac{3}{4} & \frac{1}{4} & \frac{1}{4} & \frac{3}{4} & \frac{1}{2} & \frac{1}{4}. \end{array}$$

The set $\{A \vee B, A, B\}$ is consistent, and is of size three, which is greater than the sum of $p$ over $X$,

$$\sum_{A \in X} p(A) = 2\tfrac{3}{4}$$

so $p$ and $X$ satisfies the condition of (iii). But the set $\{A \vee B, -A, -B\}$, considered by itself, violates the necessary condition expressed in (iii). It has no consistant subset of any size larger than two, and the sum of $p$ over it is $2\frac{1}{4}$.

This incompleteness of (iii) is easily remedied by requiring that every subset of $X$ also satisfy the condition.

A second way in which (iii) is incomplete is that it does not count as incoherent functions which assign low probabilities to all sentences: The assumption behind the principle is that the bettor is putting up $\sum_{A \in X} p(A)$. This incompleteness is also easily remedied, by requiring that the bettor be also willing to *take* bets, to accept $\sum_{A \in X} p(A)$ and to pay off one unit for each member of $X$ which is true. In order for such a bettor to have some chance of not losing money, there must be some possibility that he will not have to pay out any more than $\sum_{A \in X} p(A)$. In order for this to be so, there must be some sufficiently small subset $Y$ of $X$ which entails no other member of $X$, for it must be possible that he pay off just one unit for each $A \in Y$, and no more. That is to say, incorporating both of these modifications:

(iv) A necessary condition for a betting function $p$ on a finite set $X$ of sentences to be coherent is the following. For each $Y \subseteq X$, there must be some consistent subsets $Y_0$ and $Y_1$ of $Y$ such that $Y_0$ is deductively closed in $Y$ and

$$N(Y_0) \leqslant \sum_{A \in Y} p(A) \leqslant N(Y_1).$$

Coherence as formulated in this condition depends upon reference to consistency and logical closure, and is accordingly subject to the same relativization as are the concepts of transparency and probability. Without yet considering the question of a precise definition of $T$-coherence, we can formulate the condition.

(C1) Let $p$ be a betting function defined on the members of a field $\Omega$ of sentences. If $p$ is $T$-coherent, then if $X$ is any finite subset of $\Omega$, there are maximal $T$-consistent subsets $\Phi_0$ and $\Phi_1$ of $\Omega$ such that

$$N(\Phi_0 \cap X) \leqslant \sum_{A \in X} p(A) \leqslant N(\Phi_1 \cap X).$$

Before establishing and discussing results about $T$-coherence, something should be said about the difficulties which it presents as an explication of $T$-reasonableness. The most weighty of these difficulties are not so much in the notion of $T$-coherence as they are in the presuppositions required to characterize that concept. These are the difficulties in interpreting belief as willingness to bet, some of which were mentioned in II.3. There remain, however, some problems with the interpretation of reasonableness as coherence even assuming that belief is willingness to bet.

There are first several arguments that coherence is not a necessary condition for reasonableness.

### A. BELIEVING WHAT IS NECESSARY

A man whose beliefs are coherent must believe every necessary proposition to degree one and every impossible proposition to degree zero. Consider a mathematical conjecture, such as that made by Goldbach that every even number greater than 2 is the sum of two primes, or is twice some prime. No counterexample has been found to this conjecture, nor has it been proved. If it is true then it is necessary, and if false then it is inconsistent. Thus, either anyone who fails to believe it fully has incoherent beliefs, or anyone who believes it to any positive degree has incoherent beliefs. In any event, anyone who believes it to any degree distinct from both zero and one has incoherent beliefs. This shows that coherence is not necessary for reasonableness, for it is clearly not unreasonable in the present state of knowledge to have some fractional degree of belief in the conjecture, on the ground, say, that wise mathematicians seem convinced of its truth.

This argument is to some extent ameliorated by the relativization of coherence to a concept of necessity. We may explain the problem away by saying that although there are concepts of coherence for which it is incoherent to believe the conjecture to any fractional degree, there are other, weaker, concepts according to which this is not incoherent. That is so, but the generalization of this response would result in basing reasonableness upon a concept of necessity according to which whatever is necessary is known to be so. And such concepts are apparently either too weak to be of much use or too nebulous to be comprehensible in any plausible theoretical way. It would be hard, for example, to extend concepts of the first sort much beyond tautologousness or decidable fragments of predicate logic, and the applicability of coherent beliefs to matters of scientific theory would then be problematic. And, although one might argue, with respect to concepts of the second sort, that the totality of propositions known to be necessary is, if not finite, at least the union of a decidable infinite set and a large finite set, and is thus decidable, this argument does more to undermine the association of finitude and decidability than it does to establish the theoretical manageability of the notion of what is known to be necessary.

## B. RECALCITRANT BETTORS

A second argument that coherence is not necessary is that it is not plausible to require of reasonable willingness to bet that the bettor be willing to take every set of bets in accordance with his betting function. One may be willing to bet on $A$ at odds up to $\frac{1}{3}$ and on not-$A$ at odds up to $\frac{1}{3}$ and not be willing to *take* bets at $\frac{1}{3}$, that is, to bet at $\frac{2}{3}$ against either of these sentences, for the reason that the maximal odds at which one is willing to bet are a product of the amount and security of the evidence he possesses which supports the sentence in question.

This could also be taken as an argument against a presupposition of coherence, namely that one be willing to take either side of a bet.

This problem is to some extent alleviated by thinking of degree of belief as an *interval*[1] rather than as a determinate quantity. From this point of view, it is more a comment on the characteristics of willingness to bet, in particular that such characteristics do not conform to the closure and complementarity conditions required for the application of the concepts of coherence, than it is an objection to coherence as such. If the interval interpretation is correct, then some meaning should be found for coherence as applied to interval beliefs.

## C. ARGUMENTS THAT COHERENCE IS NOT SUFFICIENT FOR REASONABLENESS

A man who is willing to bet at even money that it will snow in Miami in July has not on that ground incoherent beliefs. This may even be combined with other implausible beliefs, among them beliefs about frequencies, and beliefs in propositions known to be false; he may coherently believe to degree $\frac{1}{2}$ that it snowed in Miami last July. But; it may be maintained, these beliefs are clearly unreasonable.

These cases are similar, but differ sufficiently to deserve separate consideration.

### D. COHERENCE AND CHANGE

The first of these is replied to by an argument about ways of *changing* beliefs: Very roughly put the argument is that so long as a man changes his beliefs in accordance with observation, no matter what his initial beliefs are, they will with sufficient experimentation approach the appropriate frequency, should such a frequency exist.[2] We may grant this, however, and still hold to the claim that the initial belief is unreasonable. Coherence is not offered as a criterion of reasonableness for change, but for belief at a moment. This seems to be quite right, and accordingly the question seems to be whether or not this belief by itself is reasonable, or, indeed, whether we think of it as unreasonable only because it would constitute a violation of coherence were it combined explicitly with certain other beliefs which we hold. The technique of defenders of coherence with respect to such questions is reminiscent of the technique of the utilitarians about moral decisions: You name the apparently unreasonable belief (in the case of utilitarianism, immoral act) which is not apparently incoherent (in conflict with utilitarian principles) and the coherence advocate or utilitarian will show you that it is incoherent in company with other beliefs that you hold, or that in fact utilitarian grounds can justify the decision. These techniques, in the hands of apt practitioners, are persuasive and artful, and they seem to meet many arguments of this sort.

### E. THE COHERENCE OF FALSE BELIEFS

The general form of this difficulty is easy to state: It is clearly unreasonable for a man to be willing to bet on what he knows to be false, or to bet against what he knows to be true. But unless the proposition in question is inconsistent or necessary, neither of these dispositions need be incoherent. Thus there are unreasonable bets which are not incoherent.

One might try first to meet this argument by the relativization of coherence to a concept of necessity. In order for this to work, the set $T$ of necessary transformations and truths should have to include as a 'theorem' that is as a '$T$-necessary' proposition, every proposition known to be true by the agent. There are two problems in connection with this. The first arose already in connection with mathematical beliefs; it is that the concepts of necessity which will result are theoretically unmanageable. In the present instance they should have to be, if sufficiently inclusive so as to include as necessary everything known by a given subject, pretty obviously beyond the reach of descriptive logical techniques. The second problem is that it is not easy to give a non-circular characterization of what is known by the agent, which characterization is sufficiently independent of his willingness to bet. It is difficult to think of better evidence that a man does not know *A*, than that he is willing to bet against it at some positive odds.

There is of course one sort of evidence that is relevant to this question, and this, I think, reveals an interesting feature of the concept of coherence: Not-*A* is conclusively strong evidence that the subject does not know that *A*. Thus if we know that *A* is false, we may conclude that the subject does not know that *A*. It is tempting to say also that if we know that *A*, then if the subject will bet against *A*, there is a set of bets on which he must lose, namely the bet against *A*. But this, as it stands, confuses necessity of consequence with necessity of consequent. That he will lose a bet against *A* is a necessary consequence of the proposition that we know that *A*. In order for this to imply that it is necessary that he lose the bet on *A* it should have to be necessary that we know that *A*. And such a concept of necessity, that is, one that includes all the observer's knowledge as necessary, may be a plausible base concept for coherence: It is consistent in the absolute sense, is no more bothered by the problems of mathematical beliefs than are the notions which refer to the subject's knowledge, and permits the subject to bet against *A* without undermining the evidence that *A* is necessary – such a bet would show that *he* does not know *A*, but *A* may still be necessary in the sense that we know it. But every argument that knowledge of the observer is a good basis for coherence, is an argument that coherence is a bad interpretation of reasonableness, since whether or not a man's beliefs are reasonable ought not depend upon who is considering the question. Of course, in fact, the question may be differently answered by different observers, but this does not mean that these answers may all be correct, or that the question has no precise sense independently of reference to an observer.

From these arguments and those of Chapter III section 3, we ought to conclude, it seems to me, that coherence is not an adequate explication of reasonableness. The presuppositions required are too strong and in some cases viciously unclear, and, even granting these presuppositions, there are too many important differences between our fundamental intuitions about reasonableness and the requirements of coherence, which do not seem amenable to resolution by relativization or other modification.

There is, nevertheless, an important relation between reasonableness and coherence, in some cases when the presuppositions are satisfied and a plausible concept of necessity can be fixed, reasonableness can profitably be understood as coherence. It is not easy to say in general what these cases are, but they are sufficiently many and important that the concept of coherence should not be abandoned.

This is particularly true in view of the equivalence of $T$-coherence, $T$-probability and $T$-transparency when $T$ is fixed so as to include at least all tautological logic. Of course this equivalence requires granting the presupposition of the use of coherence, namely that belief is given by willingness to bet, and thus is no answer to the objection of II.3 directed at these presuppositions. When the equivalence of probability and coherence was first established by Ramsey and, independently, De Finetti, it was thought by some to answer the question of the relation of reasonableness and probability. One of the problems raised by the advent of the measure theoretic concept of probability was that the probability of a proposition is in general, as far as the measure theoretic definition is concerned, not only not unique, but in the case of contingent propositions may be anything between zero and one inclusive. The classical theorists took probability to be unique, and they took apparent counterinstances to this to originate in unclarities and ambiguities in the notion of a possible case. La Place looked for a resolution of these ambiguities in terms of states of particles, a possible case being thought of roughly as a consistently describable arrangement of particles. Hume thought that they were resolved by interpreting probability psychologically, as an essential character of judgment, resulting from the laws of the apportionment of mental forces.

In either of these accounts, and in other similar classical accounts, the relation of probability and reasonableness is clear. Both have an ideal character, La Place's view depends upon the concept of a closed system not subject to perturbations from outside it. Hume's description of mental mechanics depends upon the clear conception of alternative outcomes each of which is subject to no extraneous mental forces. That is how apparently distinct probability assessments are to be reconciled, they are assessments of the same ideal which may differ because of the imperfect conditions under which they are made. But this option is not open in a measure theoretic development. Probability is not unique. A man's expectations, says Ramsey "May within the limits of consistency (by which he means the laws of probability) be any he likes"[3]. And this seems false of reasonableness; there are many beliefs which are probabilistic but unreasonable. It is in response to this that coherence has seemed such a powerful response. "All we have to point out", Ramsey continues, "is that if he has certain expectations he is bound in consistency to have others".[4] And indeed, though probability, which is to say coherence, may not give a unique assignment, it requires fixing only a very few partial beliefs, well selected, in order to determine the totality.

## IV.2. The Sum Condition Entails the Laws of Probability

The discussion of the preceding section treated coherence in terms of a particular set of bets, those of the form 'one if $A$, zero if not $A$'. In fact consideration of bets of this sort is also sufficient, and it can be shown that so long as certain conditions of closure and structure are satisfied, a set of bets is $T$-coherent if and only if its zero-one subset is $T$-coherent.[5] In view of this, the condition expressed in (C1) is not only necessary but sufficient, and we may take as a formulation of coherence the *finite simple sum condition*:

If $\Omega$ is a denumerable field of sentences, $T$ an absolutely consistent and at least tautological logic, and $p$ a numerical function on $\Omega$, then $\Omega$, $T$, and $p$ satisfy the *finite simple sum condition* if and only if for every finite subset $X$ of $\Omega$ there are maximal $T$-consistent $\Phi_0$ and $\Phi_1$, members of $M(\Omega)$, such that

$$N(\Phi_0 \cap X) \leqslant \sum_{A \in X} p(A) \leqslant N(\Phi_1 \cap X)$$

This condition is generalized and ramified in the sequel. As a preliminary to that discussion it will be helpful to see that the finite simple sum condition entails the laws of probability.

THEOREM 2.1. If $\Omega$ is a denumerable field of sentences, $T$ an absolutely consistent and at least tautological logic, and $p$ is a numerical function on $\Omega$, then if $\Omega$, $T$, and $p$ satisfy the finite simple sum condition, $p$ is a $T$-probability on $\Omega$.

The proof consists of a sequence of consequences of the sum condition. The arguments of $p$ are always members of $\Omega$, $\Phi_j$ ranges over the members of $M(\Omega)$, and the cardinality $N(\Phi_j \cap Y)$ is abbreviated as $N_j(Y)$.

(A1) $0 \leqslant p(A) \leqslant 1$.

*Proof*: $0 \leqslant N_j(\{A\}) \leqslant 1$, so $0 \leqslant \min_j N_j(\{A\}) \leqslant \max_j N_j(\{A\}) \leqslant 1$.

(A2) If $A$ is $T$-necessary, then $p(A) = 1$.

*Proof*: If $A$ is $T$-necessary then $A$ is a member of every maximal $T$-consistent $\Phi_j$. So, on this assumption,

$$\min_j N_j(\{A\}) = 1 \leqslant p(A)$$

and the proposition follows with the aid of (A1).

(A3) $p(A) + p(-A) = 1$.

*Proof*: $\min_j N_j(\{A, -A\}) = 1 \leqslant p(A) + p(-A) \leqslant \max_j N_j(\{A, -A\}) = 1$.

(A4) If $A$ is $T$-inconsistent, then $p(A) = 0$.

*Proof*: From the two preceding propositions.

(A5) $p$ is $T$-transparent.

*Proof*: If $A$ and $B$ are $T$-equivalent then they are members of just the same $\Phi_j$. Thus on this assumption,

$$N_j(\{A, -B\}) = N_j(\{B, -B\}) = 1$$

for each $\Phi_j$. Thus $p(A) + p(-B) = 1$; so by (A3),

$$p(A) = 1 - p(-B) = p(B).$$

(A6) $p(A \wedge B) + p(A \wedge -B) + p(-A \wedge B) + p(-A \wedge -B) = 1$.

*Proof*: $N_j(\{A \wedge B, A \wedge -B, -A \wedge B, -A \wedge -B\}) = 1$ for each $\Phi_j$.

(A7) $p(A \vee B) = p(A \wedge B) + p(A \wedge -B) + p(-A \wedge B)$.

*Proof*: By (A6) $p(A \wedge B) + p(-A \wedge B) + p(A \wedge B)$

$$= 1 - p(-A \wedge -B)$$

By (A5 and A3)

$$= p(A \vee B).$$

(A8) $p(A) = p(A \wedge B) + p(A \wedge -B)$.

*Proof*: By (A7 and (A5)

$$\begin{aligned} p(A) &= p((A \wedge \mathrm{B}) \vee (A \wedge -B)) \\ &= p((A \wedge B) \wedge (-A \vee B)) + p((A \wedge -B) \\ &\quad \wedge (-A \vee -B)) \\ &= p(A \wedge B) + p(A \wedge -B). \end{aligned}$$

(A9) If $\{A, B\}$ is $T$-inconsistent, then $p(A \vee B) = p(A) + p(B)$.

*Proof*: By (A7) and (A8),

$$p(A) + p(B) = p(A \wedge B) + p(A \wedge -B) + p(A \wedge B) + p(-A \wedge B)$$

$$p(A \vee B) = p(A \wedge B) + p(A \wedge -B) + p(-A \wedge B).$$

Since $\{A, B\}$ is $T$-inconsistent, by (A4), $p(A \wedge B)=0$, so $p(A)+p(B)= p(A \wedge -B)+p(A \wedge B)= p(A \vee B)$.

(A1), (A2), and (A9) are just the laws of finite probability. Thus theorem 2.1 is established.

## IV.3. Probability Entails the Sum Condition

The first generalization of the sum condition is to remove the restriction of finite application, that is, to allow consideration of denumerably infinite sets $X$. If $X=\{A_1, A_2, \ldots\}$ is denumerable, then three cases may arise with respect to the sum of a numerical function $p$ which assumes only non-negative values over $X$.

(i) $p$ may assign zero to all but finitely many members of $X$

(ii) Although $p$ assigns positive values to infinitely many members of $X$, the sequence of sums

(*) $p(A_1), \quad p(A_1) + p(A_2), \quad p(A_1) + p(A_2) + p(A_3), \ldots$

approaches a finite limit from below. That is to say, there is some finite $n$, such that

(a) For all $k$, $\sum_{i=1}^{k} p(A_i) \leqslant n$.

(b) If $\varepsilon$ is any positive quantity, however small, then for some $k$,

$$(n - \varepsilon) \leqslant \sum_{i=1}^{k} p(A_i).$$

In this case the sum

$$\sum_{i=1}^{k} p(A_i).$$

is said to *approach n as a limit when k increases without bound*, and this limit is identified with the sum of $p$ over $X$:

$$\sum_{A \in X} p(A) = \sum_{i=1}^{\infty} p(A_i) = \lim_{k \to \infty} \sum_{i=1}^{k} p(A_i).$$

It is useful to allow this notation in case (i) above. That is to say, if $X = \{A_1, \ldots, A_r\}$ then we allow that

$$\lim_{k\to\infty} \sum_{i=1}^{k} p(A_i) = \sum_{i-1}^{r} p(A_i).$$

It may be remarked that finite sets $X$ are included in the description of case (i).

In the third case

(iii) The sequence of sums (*) approaches no finite limit when $k$ increases without bound. That is to say that for every finite $n$, there is some $k$, such that

$$n < \sum_{i=1}^{k} p(A_i).$$

In this case the sum of $p$ over $X$ is said to be infinite.

The formulation of the simple sum condition to cover these cases is straightforward:

Simple sum condition. If $\Omega$ is a denumerable field of sentences, $T$ an absolutely consistent and at least tautological logic, and $p$ a numerical function on $\Omega$, then $\Omega$, $T$, and $p$ satisfy the *simple sum condition* if and only if for every subset $X$ of $\Omega$,

(i) If the sum of $p$ over $X$ is finite, then there are maximal $T$-consistent $\Phi_0$ and $\Phi_1 \in M(\Omega)$ such that

$$N(\Phi_0 \cap X) \leqslant \sum_{A \in X} p(A) \leqslant N(\Phi_1 \cap X).$$

(ii) If the sum of $p$ over $X$ is infinite, then there is no finite upper bound on the quantities $N_j(X)$. I.e., for every finite $n$ there is some $\Phi_j \in M(\Omega)$ such that $n < N_j(X)$.

In much of what follows we shall discuss probabilities defined on fields $\Omega$ and based on logics $T$. It will simplify that discussion to introduce the unifying concept of a *probability system*.

The triple $\langle \Omega, T, p \rangle$ is a probability system if and only if

(i) $\Omega$ is a finite or denumerably infinite field of sentences.

(ii) $T$ is an absolutely consistent and at least tautological logic defined on $\Omega$.

(iii) $p$ is a $T$-probability measure on $\Omega$.

In the sequel characteristics of $\Omega$, $T$ and $p$ may be applied to the system $\langle \Omega, T, p \rangle$ when the sense is obvious. Thus, for example, $\langle \Omega, T, p \rangle$ may be said to have a finite base, to be compact, or to satisfy the simple sum condition meaning that $T$ inconsistency is compact in $\Omega$, that $\Omega$ has a finite $T$-base, or that $\Omega$, $T$ and $p$ satisfy the simple sum condition.

The main theorem of this section, which is proved by a succession of special cases may now be simply stated.

THEOREM 3.3. If $\langle\Omega, T, p\rangle$ is a compact probability system, then $\langle\Omega, T, p\rangle$ satisfies the simple sum condition.

We recall that a *constitution* (II.1) of a subset $Y$ of $\Omega$ is a set which includes for each sentence $A \in Y$, either $A$ or its negation. And that a subset $Y$ of $\Omega$ is a *finite T-base* for $\Omega$, if (i) $Y$ is finite (ii) Every constitution of $Y$ is $T$-consistent (iii) If $A$ is any sentence in $\Omega$, then there is a set $C_A$ of constitutions of $Y$ such that $A$ is $T$-equivalent to the disjunction of the conjunctions $\wedge Y_i$ for $Y_i \in C_A$.

The first special case to consider is

THEOREM 3.1. If $\langle\Omega, T, p\rangle$ is a compact probability system and has also a finite base, then $\langle\Omega, T, p\rangle$ satisfies the simple sum condition as far as finite subsets of $\Omega$ are concerned.

*Proof of theorem* 3.1: Since $\Omega$ has a finite $T$-base, by theorem III.1.4, each maximal $T$-consistent $\Phi_j$ can be given as the set of $T$-consequences of some constitution of the finite $T$-base $Y$, and this correspondence is one to one. Thus, writing '$\lesssim$' to stand for $T$-implication, the maximal $T$-consistent $\Phi_j$ and the constitutions $Y_j$ of the base match up so that

$$\Phi_j = \{A \mid Y_j \lesssim A\} \cap \Omega$$

for each index $j$. Thus for each $\Phi_j \in M(\Omega)$ we have that

$$N_j(X) = N(\Phi_j \cap X) = N(\{A \mid Y_j \lesssim A\} \cap X)$$

for each $X \subseteq \Omega$. And to establish the sum condition it will suffice to show that for each $X \subseteq \Omega$ there are constitutions $Y_0$ and $Y_1$ of the base $Y$ such that

$$N(\{A \mid Y_0 \lesssim A\} \cap X) \leqslant \sum_{A \in X} p(A) \leqslant N(\{A \mid Y_1 \lesssim A\} \cap X).$$

For each constitution $Y_j$ we can define the function $p_j$ on $\Omega$:

$$p_j(A) = 1 \Leftrightarrow Y_j \lesssim A$$
$$p_j(A) = 0 \text{ otherwise.}$$

Each such $p_j$ is a two valued $T$-probability on $\Omega$. To establish the sum condition it will suffice to show that there are among these functions probabilities $p_0$ and $p_1$ such that

$$\sum_{A \in X} p_0(A) \leqslant \sum_{A \in X} p(A) \leqslant \sum_{A \in X} p_1(A)$$

for the sum of such a function over the set $X$ is just the number of members of $X$ each of which is entailed by the constitution of $Y$ to which the function corresponds.

By Theorem III.3.2, if $A$ is a $T$-consistent proposition in $\Omega$, then

$$p(A) = \sum_{Y_j \lesssim A} p(\wedge Y_j)$$

so for each $A \in \Omega$,

$$p(A) = \sum_j [p_j(A) \cdot p(\wedge Y_j)]$$

where $j$ ranges over *all* the constitutions of $Y$, for the measure $p_j$ assigns zero to $A$ if $Y_j$ does not imply $A$, and one to $A$ if $Y_j$ does imply $A$. So for $X \subseteq \Omega$

$$\sum_{A \in X} p(A) = \sum_{A \in X} \sum_j [p_j(A) \cdot p(\wedge Y_j)]$$

Factoring this sum we obtain

$$\sum_{A \in X} p(A) = \sum_j [p(\wedge Y_j) \cdot \sum_{A \in X} p_j(A)].$$

The import of the last equation is that the sum of $p$ over $X$ is a weighted average of the various sums

$$\sum_{A \in X} p_j(A).$$

Since the weights in this average are the quantities $p(\wedge Y_j)$, each of which is between zero and one inclusive, the average can neither exceed nor be exceeded by all of the above sums. In view of the relations among the various $p_j$ and $p$, the sum condition follows. This completes the argument for theorem 3.1.

The next step is to remove the requirement that the system have a finite base, continuing, however, the restriction to finite subsets $X$ of $\Omega$.

THEOREM 3.2. If $\langle\Omega, T, p\rangle$ is a compact probability system, then $\langle\Omega, T, p\rangle$ satisfies the simple sum condition as far as finite subsets $X$ of $\Omega$ are concerned.

*Proof of theorem* 3.2. Let $X$ be a finite subset of $\Omega$. $X$ generates a field $\Omega_X$, which is equivalently the intersection of all fields containing $X$, and the result of closing $X$ under finite truth-functions. This field will have only finitely many (at most $2^k$ if $X$ has $k$ members) atoms, or strongest $T$-consistent members. These correspond to the maximal $T$-consistent sets of the Tarski-Lindenbaum Algebra. By adding finitely many propositions to $\Omega_X$ we can increase the number of these atoms to some finite power of 2, say $2^m$. The logical closure of the result is a field $\Omega_X^*$ with exactly $2^m$ atoms. Such a field has a finite $T$-base of size $m$. Now the original field may be extended to a field $\Omega^*$ of which $\Omega_X^*$ is a subfield, and the measure $p$ may be extended to $\Omega^*$ without affecting its definition on $X$.

Since $X$ is a subset of the field $\Omega_X^*$ with a finite $T$-base, $\langle\Omega_X^*, T, p\rangle$ satisfies the conditions of theorem 3.1. That theorem entails that the sum condition holds for $X$.

We turn now to the support of theorem 3.3 with no restrictions.

*Proof of theorem* 3.3: Assume the subset $X$ of $\Omega$ to be denumerably infinite. Let

$$X = \{A_1, A_2, \ldots\}.$$

We define a sequence of finite subsets of $X$. Let

$$X_0 = \Lambda$$

and for each $i > 0$, let

$$X_i = \{A_1, \ldots, A_i\}.$$

We call the sequence $\{X_i\}$ a *covering nest* for $X$. (Such sequences are also used in the definitions and arguments of the next chapter.) Each member of the nest is finite, and the union of the nest is the set $X$.

We consider two cases of theorem 3.3 according to whether the sum of $p$ over $X$ is finite or infinite.

3.3 FINITE CASE. $\sum_{A \in X} p(A) = m$ for some finite $m$.

In this case the theorem asserts lower and upper bounds on the finite quantity $m$. $m$ must be shown to lie between the minimum and the maximum respectively of the quantities $N_j(X)$ as $j$ varies, indexing the various maximal $T$-consistent subsets $\Phi_j$ of $\Omega$.

We argue first to the existence of a lower bound, of a $\Phi_0$ as described in the theorem. First some lemmas in which some consequences of finiteness and of the compactness assumption are developed.

LEMMA A. In the finite case, for some $\Phi_j \in M(\Omega)$ the quantity $N_j(X)$ is finite.

*Proof of lemma A*: If this quantity is never finite then as $i$ increases without bound, indexing increasingly larger members of the nest $\{X_i\}$, the quantity

$$\min_j N_j(X_i)$$

also increases without bound. Thus, on this assumption, for some $i$

$$m < \min_j N_j(X_i)$$

Since $X_i \subseteq X$, $\sum_{A \in X_i} p(A) \leqslant m$. Thus, we should have, in contradiction of theorem 3.2, that for some finite $X_i$

$$\sum_{A \in X_i} p(A) \leqslant m < \min_j N_j(X_i)$$

and the lemma follows by avoiding this contradiction.

LEMMA B. If for every maximal $T$-consistent $\Phi_j$ the intersection of $X$ with $\Phi_j$ is not null, then $X$ has a finite subset with this characteristic, and, in particular, for some $X_i \in \{X_i\}$, for every index $j$, $N_j(X_i) > 0$.

*Proof of lemma B*: If for every $j$, $X \cap \Phi_j \neq \Lambda$, then for every $j$, there is some $A \in X \cap \Phi_j$, and for every $j$ there is some $A$ such that $A \in X$ and $-A$ is not a member of $\Phi_j$. Thus the set $\{-A \mid A \in X\}$ is a subset of no $\Phi_j$, and is hence not $T$-consistent. By the assumed compactness of $T$ in $\Omega$, this set has a finite $T$-inconsistent subset, call it $W$. $W$ is finite, hence for some $i$,

$$W \subseteq \{-A \mid A \in X_i\}$$

so the set $\{-A \mid A \in X_i\}$ is $T$-inconsistent, and is thus a subset of no $\Phi_j$. Since the sets $\Phi_j$ are maximal, for each $\Phi_j$, some $A \in X_i$ is also a member of $\Phi_j$. Thus for each $\Phi_j$, $N_j(X_i) > 0$.

LEMMA C. If $\min_j N_j(X)=n$, then for some $X_i \in \{X_i\}$, $\min_j N_j(X_i)=n$.

*Proof of lemma C*: Form $X^{(n)}$, the set of all $n$-membered conjunctions of elements of $X$. The hypothesis entails that $X^{(n)}$ intersects every $\Phi_j$. Thus, by lemma B, for some $X_i$, the set of $n$-membered conjunctions of elements of $X_i$; $X_i^{(n)}$, intersects every $\Phi_j$. Hence, for some $X_i$, for every $\Phi_j$, $N_j(X_i) \geqslant n$. Thus for this $X_i$, $\min_j N_j(X_i) \geqslant n$. Further,

$$\min_j N_j(X_i) \leqslant \min_j N_j(X) = n$$

which completes the argument for lemma C.

The proof of the lower bound part of the finite case is now quite simple: We are assuming that the sum of $p$ over $X$ is the finite quantity $m$. Thus by lemma A, for some $j$, $N_j(X)$ is finite. Let

$$\min_j N_j(X) = n.$$

Then by lemma C, for some $i$,

$$\min_j N_j(X_i) = n.$$

Now if $m < n$, then

$$\sum_{A \in X_i} p(A) < \sum_{A \in X} p(A) = m < n = \min_j N_j(X_i)$$

contradicting theorem 3.2. This establishes the existence of a $\Phi_0$ as asserted in theorem 3.3, in the case in which the sum of $p$ over $X$ is finite.

To complete this case it remains only to establish the existence of a $\Phi_1$ determining an upper bound on $\sum_{A \in X} p(A)$. The argument is brief and straightforward: Suppose in contradiction to the theorem that

$$\max_j N_j(X) < \sum_{A \in X} p(A).$$

Then

$$\max_j N_j(X) < \lim_{k \to \infty} \sum_{i=i}^{k} p(A_i).$$

Since the finite sum of $p$ over $X$ is just this limit. Thus for some $k$,

$$\max_j N_j(X_i) \leqslant \max_j N_j(X) < \sum_{i=i}^{k} p(A_i).$$

in contradiction of theorem 3.2. This completes the argument for the case in which the sum of $p$ over $X$ is finite.

*Proof of theorem 3.3 case* (*ii*): In this case $\sum_{i=i}^{k} p(A_i)$ increases without bound as $k$ increases without bound, and the theorem asserts that for each finite $n$, there is some $\Phi_j \in M(\Omega)$ such that $n < N_j(X)$. To see the truth of this, assume it false. Then for some $n$, for all $\Phi_j$, $N_j(X) < n$. By the assumption of this case, for some $k$, $n < \sum_{i=i}^{k} p(A_i)$. And since $N_j(X_k) \leqslant N_j(X) < n$ for all $\Phi_j$, we have that for some $k$

$$\max_j N_j(X_k) < \sum_{i=1}^{k} p(A_i) = \sum_{A \in X_k} p(A)$$

in contradiction of theorem 3.2. This completes the proof of theorem 3.3.

## NOTES

[1] Cf. *Koopman* and *Good*. Henry Kyburg has also done some work with interval interpretations of probability, but as far as I know, it rests unpublished.
[2] Cf. Ramsey 'Truth and Probability' in *Ramsey* (103L), and the discussion in *De Finetti*, chapter VI.
[3] P. 189.
[4] See also *De Finetti*, Chapter VI.
[5] See *Suppes and Zinnes*, Theorem 15, and the discussion of Chapter II section 2 above.

CHAPTER V

# PROBABILITY AND INFINITY

## V.1. Subjectivism and Infinity

In the preceding chapter the relations between the epistemic concept of probability, as a logic of partial belief, and the measure-theoretic concept of probability were developed for objects – sentences – of finite complexity. The account showed that there are nice relations among (i) probabilities on sets of sentences, (ii) probabilities on the Tarski-Lindenbaum Algebras associated with those sets of sentences, (iii) the transparency of partial belief, and (iv) a condition formulated there, based upon the concept of coherence, called the simple sum condition. That account seems satisfactory as far as it goes, but, leaving aside the question to what extent it can be made less relativistic, that is to say, in what ways appropriate logics $T$ can be more precisely specified, it remains obviously incomplete in one important formal respect: The functions there defined apply only to finitely complex objects, and, in particular, are additive only over finite disjunctions of incompatible sentences, or over finite unions of disjoint sets. Indeed, the closure properties of Boolean Algebras are finite too; if $\alpha_i, \cdots, \alpha_k$ are members of a Boolean Algebra of sets then the finite union

$$\bigcup_{i=1}^{k} \{\alpha_i\} = \alpha_1 \cup \cdots \cup \alpha_k$$

is also a member of that algebra, and if $A_1, \ldots, A_k$ are members of a field of sentences, then the finite disjunction

$$\bigvee_{i=1}^{k} \{A_i\} = A_1 \vee \cdots \vee A_k$$

is also a sentence in that field. But a Boolean Algebra may include a denumerable infinity of distinct sets and fail to include their union, and a field of sentences may include a denumerable infinity of distinct sentences and fail to include their disjunction. It being, in this latter instance, perhaps not even clear what the disjunction of an infinity of distinct sentences is.

In the standard measure-theoretic account of probability, however, probability measures are defined not over Boolean Algebras in general, but over sigma algebras, which include denumerably infinite as well as finite unions of their members.[1] And probabilities are denumerably as well as finitely additive.

If $\mathfrak{B}$ is a sigma algebra of subsets of $M(A)$ then a *probability* on $\mathfrak{B}$ is a function $\mu$ on $\mathfrak{B}$ such that

(i) $0 \leqslant \mu(\alpha) \leqslant 1$.

(ii) $\mu(M(A)) = 1$.

(iii) If $C$ is a denumerable subcollection of $A$, no two members of which have any common element, then

$$\mu[\cup C] = \sum_{\alpha \in C} \mu(\alpha).$$

The third of these conditions is of course not applicable to Boolean Algebras in general, and this raises the interpretative question of the resolution of this incongruity: The logical structure of sentences seems to be in general finite, but probabilities are defined on objects of denumerably infinite complexity. Let us discuss some reactions to this situation.

It may be insisted that propositions or sentences, the objects of belief, are finite. That is to say, that there are no infinite conjunctions or disjunctions, and that the probability of propositions is finitely additive, but not denumerably additive. Thus one would reject the persuasive definition of probabilities as denumerably additive normal measures, and count as probabilities just the finitely additive normal measures on Boolean Algebras.

This point of view particularly characterizes subjectivistic approaches to the foundations of probability, according to which probability is intimately, if not essentially, associated with betting and coherence. From such a point of view it is not at all clear what empirical meaning can be given to the probability of an infinite disjunction, since it is not clear what sense can be given to bets on infinite totalities (which are of course to be distinguished from infinite collections of bets). One finds a certain nominalism here, too: The preference is to avoid as much as possible attributing such operations as the taking of limits to the mind of the believer: It is clear that for subjectivism to function we must think of the believer as computing, but from an empiricistic and nominalistic viewpoint, we should take care, first, to assure that we limit such assumptions to as few and as simple operations as possible, and, second, we should be sure that we can consistently think of the believer as failing to compute correctly: of, for example, failing to believe in the disjunction of incompatible propositions to an extent which equals the sum of the extents to which he believes in the propositions. The notion of coherence has no application if this is not so, and we can make no distinction between belief *simpliciter* and probabilistic belief. Hume has given the model for such theories: It has the mind combining and summing forces of beliefs in incompatible propositions which must be held before the mind, from which it follows, by the Humean analogy of thought and vision, that there must be only finitely many of them. As far as the notion of incompatibility is concerned, Hume's view is that the believer has no genuine view of incompatibility, and that what we mean by incompatibility of objects (propositions) is really just the psychological incompatibility of the ideas of those objects. Subjectivistic views, of which Hume's may be considered the prototype, make no provision for a man failing to hold incompatible what are really incompatible propositions, for the only notion of incompatibility provided on Hume's view is just so holding, or formal incompatibility, and in the views of his successors, this is either not questioned, or beliefs are assumed to be transparent for some, usually implicit, notion of logical incompatibility.

Now the ways in which such views must consider questions of denumerable additivity are clear. We must either ignore denumerable additivity altogether as embodying the psychological impossibility of confronting infinitely many distinct alternatives, or we must treat it as an *a priori* concept, as incompatibility is sometimes treated, and hold that the constraint of denumerable additivity is never violated, since it expresses a defining characteristic of belief in infinitely complex propositions. The latter option is not genuine for a subjectivistic theory, because it would leave mysterious the connection between beliefs in infinite disjunctions, and beliefs in their finite subsets, and it would raise the question why infinite additivity should be assumed an *a priori* law of thought, no error of computation being possible on the part of the believer, while an *a posteriori* account is given of the apparently simpler computations involved in finite additivity. It would be at least strange to assume believers to be infallible computers of infinite sums, but to allow them the possibility of error in the case of finite sums. This would be all the more strange since the logic of finite addition can be completely specified in first-order arithmetic, while that of limits and convergence requires at least a theory of the second order, and, indeed, one for which the domains of the second order variables are still not completely understood.

## V.2. Extensions of probabilities

In the present section we discuss a method of extending fields of sentences so as to allow the extension of probabilities defined on them to probabilities on objects of denumerable complexity. This is done by extending the base logics to define logical relations among *sets* of sentences and then by extending the probabilities defined on the members of a field of sentences to assign values also to subsets of the field. These extensions turn out to be simply and easily characterizable.

As far as the extension of the logic is concerned, the task is mainly terminological: We have already in chapter III defined $T$-implication and related logical notions among sets of sentences. Here it is assumed in particular that if $\Omega$ is a denumerable field of sentences, and $T$ is absolutely consistent and at least tautological, then, where $A$ is any member of $\Omega$, $X$ and $Y$ are subsets of $\Omega$, and $\Phi$ ranges over the maximal $T$-consistent subsets of $\Omega$;

| | | |
|---|---|---|
| $X$ is $T$-consistent | $\Leftrightarrow$ | $X \subseteq \Phi$, for some $\Phi$ otherwise $T$-inconsistent |
| $X$ $T$-implies $A$ | $\Leftrightarrow$ | $X \cup \{-A\}$ is $T$-inconsistent |
| $X$ $T$-implies $Y$ | $\Leftrightarrow$ | $X$ $T$-implies every member of $Y$ |
| $X$ and $Y$ are $T$-equivalent | $\Leftrightarrow$ | Each $T$-implies the other |
| $X$ is $T$-necessary | $\Leftrightarrow$ | $X \subseteq \Phi$, for every $\Phi$. |

Some *disjunctive* concepts are also employed, although the terms *necessity*, *equivalence*; and *implication* are avoided in view of the possibility of confusion with the more customary conjunctive concepts: In particular;

| | |
|---|---|
| $X \cap \Phi \neq \Lambda$ for every $\Phi$ | functions like necessity |
| $X \cap \Phi \neq \Lambda$ for some $\Phi$ | functions like possibility |
| $X \cap \Phi \neq \Lambda \Rightarrow Y \cap \Phi \neq \Lambda$ | functions like implication. |

Each member of a field of sentences has a finite logical structure. There is thus no clear way to apply the requirements of denumerable additivity to probability measures on fields of sentences. One fairly obvious way to provide an opportunity for the application of this requirement is to think of probability as applying to sets of sentences, disjunctively considered: To think of the probability of a set $X$ of sentences as the probability that some sentence in the set is true. Then, clearly, if $X = \{A_1 \ldots\}$ is a denumerable set of pairwise incompatible sentences, so that no two of them can both be true, then the probability of $X$ disjunctively considered should be the sum of the probabilities of the members (or the distinct unit subsets) of $X$. That is a simple way to apply the constraint of denumerable additivity.

Now let us consider the question of the possibility of plausibly defining probability disjunctively considered, and let us put the question in this form: Given a probability system $\langle \Omega, T, p \rangle$, how should $p$ be extended to a disjunctive $T$-probability on the subsets of $\Omega$, where, intuitively, the disjunctive probability of a set $X$ is the probability that some sentence in $X$ is true?

A part of this question has an obvious answer: If $\check{p}$ is the disjunctive $T$-probability to which $p$ is extended, then if $X$ is a finite non-null subset of $\Omega$, the disjunctive probability of $X$, $\check{p}\,(X)$, should be just the probability of the disjunction of the members of $X$, $p(\vee X)$. That is the only definition which assures additivity in the finite case. Another fairly obvious requirement is this; the disjunctive probability of the null set should be zero. These two partial definitions have the following consequences, where $X$ and $Y$ range over finite subsets of $\Omega$.

(1) $\check{p}$ is transparent for $T$-equivalence of the disjunctions of its arguments. I.e., if $\vee X$ is $T$-equivalent to $\vee Y$, then $\check{p}(X) = \check{p}(Y)$.

(2) $\check{p}$ is monotone non-decreasing for the subset relation. If $X \subseteq Y$ then $\check{p}(X) \leqslant \check{p}(Y)$.

(3) If $X$ intersects every maximal $T$-consistent $\Phi$, then $\check{p}(X) = 1$.

(4) If $A \in \Omega$ then $\check{p}(\{A\}) = p(A)$.

(5) If the members of the subset $Z$ of $\Omega$ are pairwise $T$-incompatible, then if $X$ and $Y$ are subsets of $Z$, $\check{p}(X \cup Y) = \check{p}(X) + \check{p}(Y)$. So, for example; $\check{p}(\{A, -A\}) = 1 = p(A) + p(-A)$. And consequently, if $A$ is any member of $\Omega$, $X$ is a finite subset of $\Omega$, and $X_1, \ldots, X_k$ the constitutions of $X$:

$$p(A) = \sum_{i=1}^{k} p(A \wedge \wedge X_i) = \check{p}(\{A \wedge \wedge X_i\}).$$

So, for example,

$$\begin{aligned} p(A) &= \check{p}(\{A\}) = p(A \wedge B) + p(A \wedge -B) \\ &= \check{p}(\{A \wedge B, A \wedge -B\}). \end{aligned}$$

So far, of course, we have not really extended the measure $p$ on sentences.

Except for the definition of $p$ on the null set, its values are just those given by $p$ to the sentences most obviously associated with the sets for which it is defined. The interesting question is that of the extension of $\check{p}$ to denumerably infinite sets of sentences. And the answer to that question is also not hard to find, at least not if we attend to the requirements of additivity which it should satisfy. It is to take $\check{p}(X)$ to be the least upper bound of the values of $\check{p}$ for the finite subsets of $X$.

$$\begin{aligned}\check{p}(X) &= \sup \{\check{p}(X_i) \mid X_i \text{ a finite subset of } X\} \\ &= \sup \{p(\vee X_i) \mid X_i \text{ a finite subset of } X\}.\end{aligned}$$

It is, first, clear that this definition is consistent with the proposal for finite sets. For if $X$ is a finite set, then the collection of finite subsets of $X$ is finite, and the least upper bound of the values assigned by $\check{p}$ to members of this collection is just the maximum of the values assigned to members of the collection. In view of the monotonicity of $\check{p}$ with the superset relation, this is just the value of $\check{p}(X)$. Thus the definition in terms of upper bounds has the earlier proposal as a consequence. Further, since it is assumed that $p$ is a $T$-probability, $0 \leqslant p(\vee X_i) \leqslant 1$ for every finite subset $X_i$ of $\Omega$ and thus the least upper bound $\sup \{\check{p}(X_i) \mid X_i \text{ a finite subset of } X\}$ will always exist. Things can thus be simplified, and the discussion so far summarized, in the following definition: If $\langle \Omega, T, p \rangle$ is a probability system, then the *disjunctive extension* of $p$ to subsets of $\Omega$ is the function $\check{p}$ on subsets of $\Omega$ defined

$$\begin{aligned}&\text{If } X \neq \wedge \text{ then } \check{p}(X) = \sup \{p(\vee X_i) \mid X_i \text{ a finite subset of } X\} \\ &\check{p}(\wedge) = 0.\end{aligned}$$

The justification of this definition will be in its consequences. We postpone these derivations, however, to turn to the parallel development of the dual notion suggested by the preceding discussion, that of the *conjunctive* probability of a set of sentences. The considerations leading up to the definition are quite analogous to those developed with respect to disjunctive probability, so we state the definition directly and then discuss some consequences.

If $\langle \Omega, T, p \rangle$ is a probability system, then the *conjunctive extension* of $p$ to subsets of $\Omega$ is the function $\hat{p}$ on subsets $X$ of $\Omega$ defined.

$$\begin{aligned}&\text{If } X \neq \wedge \text{ then } \hat{p}(X) = \inf \{p(\wedge X_i) \mid X_i \text{ a finite subset of } X\} \\ &\check{p}(\wedge) = 1.\end{aligned}$$

Since $p(\wedge X_i)$ is never less than zero, the greatest lower bound of any collection $\{p(\wedge X_i)\}$ always exists, and thus $\hat{p}$ is well defined.

Notice also that since $\wedge\{A\}=\vee\{A\}=A$ for individual sentences $A$, $\hat{p}$ and $\check{p}$ agree with each other for unit sets; both just take the value of $p$ for the sentence in the set. We have the following characteristics of $\hat{p}$ for finite subsets $X$ and $Y$ of $\Omega$:

(1) $\hat{p}$ is $T$-transparent for $T$-equivalence of finite sets.

(2) $\hat{p}$ is monotone non-increasing for the subset relation among finite sets; If $X \subseteq Y$ then $\hat{p}(X) \geqslant \hat{p}(Y)$.

(3) If $X$ is $T$-necessary then $\hat{p}(X)=1$.

The conjunctive and disjunctive extensions of a given probability $p$ may be intuitively related by considering, instead of the collection of all finite subsets of a given set $X$, a particular *covering nest*, well-ordered by the subset relation. Let $X=\{A_1, \ldots\}$ be a subset of a denumerable field on which the $T$-probability $p$ is defined, and let $\hat{p}$ and $\check{p}$ be the conjunctive and disjunctive extensions of $p$. Define the covering nest $\{X_i\}$

$$X_0 = \Lambda$$
$$X_{n+1} = X_n \cup \{A_{n+1}\}$$

so

$$X_0 \subseteq X_1 \subseteq \cdots \subseteq X.$$

Every $X_i$ is finite

$$\bigcup_i \{X_i\} = X$$

and $X$ is the set-theoretic least upper bound of the covering nest $\{X_i\}$. Also, if $Y$ is any finite subset of $X$, then there are $X_i$ and $X_j$, members of $\{X_i\}$, such that $X_i \subseteq Y \subseteq X_j$.

In view of the monotonicity properties of $\hat{p}$ and $\check{p}$, as the index $i$ increases, $\hat{p}(X_i)$ is non-increasing and $\check{p}(X_i)$ is non-decreasing. Thus if $Y$ is any finite subset of $X$, there are $i$ and $j$ such that

$$\hat{p}(X_i) \geqslant \hat{p}(Y) \geqslant \hat{p}(X_j)$$
$$\check{p}(X_i) \leqslant \check{p}(Y) \leqslant \check{p}(X_j)$$

and consequently the least upper bound of $\check{p}$ over all finite subsets of $X$ is the least upper bound over the nest $\{X_i\}$, and analogously for the greatest lower bound of $\hat{p}$.

$$\check{p}(X) = \sup \{p(\vee X_i) \mid X_i \mathrel{\varepsilon} \text{the nest } \{X_i\}\}$$
$$\hat{p}(X) = \inf \{p(\wedge X_i) \mid X_i \mathrel{\varepsilon} \text{the nest } \{X_i\}\}.$$

An important consequence of the first definition is the denumerable additivity of $\check{p}$ in the sense that

LEMMA A. If $\langle \Omega, T, p \rangle$ is a probability system and $\check{p}$ is the disjunctive extension of $p$, then if $X = \{A_1, \ldots\}$ is a denumerable set of pairwise $T$-incompatible members of $\Omega$,

$$\check{p}(X) = \sum_{A_i \in X} p(A_i).$$

*Proof:* Consider the nest $\{X_i\}$ as defined above. Since $p$ is finitely additive over finite disjunctions of pairwise $T$-incompatible sentences, for each $n$

$$p(\vee X_n) = \sum_{i=1}^{n} p(A_i).$$

We need first to see that as $n$ increases without bound, both sides of this equation approach a limit. Consider the left side. If it does not approach a limit no greater than one, then for some $n$, $\sum_{i=1}^{n} p(A_i) > 1$. But this violates the assumption that $p$ is a probability. So clearly both sides approach limits, which are the same.

$$\lim_{n \to \infty} p(\vee X_n) = \lim_{n \to \infty} \sum_{i=1}^{n} p(A_i).$$

The least upper bound of the values assigned by $p$ over the nest $\{X_i\}$ is just the left side of the above equation

$$\sup \{p(\vee X_i)\} = \lim_{n \to \infty} \sum_{i=1}^{n} p(A_i)$$

and, as was remarked above, this is $\check{p}(X)$, so

$$\check{p}(X) = \lim_{n \to \infty} \sum_{i=1}^{n} p(A_i)$$

and since the limit of the sum exists, it follows that the sum converges, so

$$\check{p}(X) = \sum_{i=1}^{\infty} p(A_i) = \sum_{A_i \in X} p(A_i)$$

which establishes lemma A.

In many interesting cases the field $\Omega$ will also be a *first-order language*, built up from a finite collection of predicates and a denumerable collection of individual constants, by means of truth-functional connectives and the devices of quantification. If this is so, and if in addition $T$ is an extension of classical predicate logic as regards the sentences in $\Omega$, then $\langle\Omega, T, p\rangle$ is said to be *predicative*. In this case the class of $T$-necessary sentences of $\Omega$ will constitute a first-order theory in the usual sense of that term.

The preceding discussion and some additional results are conveniently summarized in a six-part theorem.

THEOREM 2.1 Let $\langle\Omega, T, p\rangle$ be a probability system, let $X$ range over subsets of $\Omega$, and for (4a) and (5b) assume that $T$-inconsistency is compact in $\Omega$. Then

(1a) $\hat{p}\{-A \mid \mathrm{A} \in X\} = 1 - \check{p}(X)$

(1b) $\check{p}\{-A \mid A \in X\} = 1 - \hat{p}(X)$.

*Proof:* (1a) holds for finite sets $X$ in virtue of $p$ being a $T$-probability. Thus if $\{X_i\}$ is a covering nest of $X$

$$p(\wedge\{-A \mid A \in X_i) = 1 - p(\vee X_i)$$

for each $i$. (1a) follows by taking the infimum over $i$ for each side. The proof of (1b) is quite symmetrical.

(2a) $\hat{p}(\wedge) = 1$

(2b) $\check{p}(\wedge) = 0$.

(3) If $X$ is not null then $0 \leqslant \hat{p}(X) \leqslant \check{p}(X) \leqslant 1$.

*Proof:* It has already been remarked that both extensions are bounded by zero and one. To see the inequality between them; on the assumption that $X$ is not null; notice that because of the monotonicity of $p$ with $T$-implication among sentences of $\Omega$,

$$p(\wedge X_i) \leqslant p(\vee X_i)$$

for each non-null finite $X_i \subseteq X$. Thus

$$\hat{p}(X) = \inf\{p(\wedge X_i)\} < \sup\{p(\vee X_i)\} = \check{p}(X)$$

where $X_i$ ranges over the finite subsets of $X$.

(4a) On the assumption that $T$-inconsistency is compact in $\Omega$, if $X$ is $T$-inconsistent, then $\hat{p}(X) = 0$.

(4b) Without the assumption of compactness, if $X$ intersects no maximal $T$-consistent subset of $\Omega$ then $\check{p}(X) = 0$.

*Proof* (4a): Assume $X$ to be $T$-inconsistent. By compactness, some finite subset $X_i$ of $X$ is $T$-inconsistent. Thus the measure $p$ assigns zero to $\wedge X_i$ for some finite $X_i \subseteq X$, and hence the infimum of the values assigned by $\hat{p}$ to the finite subsets of $X$ cannot exceed zero. (4a) follows by (3).

*Proof* (4b): If $X$ intersects no maximal $T$-consistent subset of $\Omega$ then none of its finite subsets can do so. Thus every member of $X$ is $T$-inconsistent, and every finite disjunction of members of $X$ is $T$-inconsistent. Thus since $p$ is a $T$-probability

$$p(\vee X_i) = 0$$

for each finite $X_i \subseteq X$, and hence the supremum of these values – again employing (3) – is just zero.

(5a) Without the assumption of compactness, if $X$ is $T$-necessary, then $\hat{p}(X)=1$.

(5b) On the assumption that $T$-inconsistency is compact in $\Omega$, if $X$ intersects every maximal $T$-consistent subset of $\Omega$ then $\check{p}(X)=1$.

*Proofs:* For (5a) use (1b) and (4b). For (5b) use (1a) and (4a).

(6) If the members of $X$ are pairwise $T$-incompatible, then

$$\check{p}(X) = \sum_{A \in X} p(A).$$

*Proof:* Lemma A.

In view of this theorem we shall, in the sequel, speak of the conjunctive and disjunctive extensions of a $T$-probability as themselves $T$-probabilities.

## V.3. Independence

The interest of disjunctive probability is in additivity. One must be able to sum the probabilities of incompatible alternatives to find the probability that one of them will occur. *Incompatibility* is itself a purely logical property, given a logic $T$ it is established what sentences are pair-wise incompatible; and does not depend upon probabilistic concepts. The constraint of additivity is thus a constraint placed upon probability by a concept which is itself non-probabilistic. The crucial concept for conjunctive probability, on the other hand, does not have this characteristic. What is important for conjunctive probability is *independence*: Incompatible alternatives affect each other completely; should one occur, the other cannot occur. Independent events affect each other not at all, the chance of either occurring is the same whether or not the other occurs. As this remark suggests, the concept of independence is not, in contrast to that of incompatibility, extra-probabilistic. Or, to put this another way, events, or the sentences which express them, may be independent in one probability function but not in another. The simplest definition of independence is this:

Two sentences $A$ and $B$ are *independent* in a measure $p$ if

$$p(A \wedge B) = p(A) \cdot p(B).$$

If neither $p(A)$ nor $p(B)$ is zero, then $A$ and $B$ are independent in $p$ if and only if

$$\frac{p(A \wedge B)}{p(A)} = p(B)$$

which in this case is equivalent to

$$\frac{p(A \wedge B)}{p(B)} = p(A)$$

From which it follows that when $p$ is zero at neither $A$ nor $B$ they are independent in $p$ just in case

$$p_A(B) = p(B)$$

and also just in case

$$p_B(A) = p(A).$$

Independence in a $T$-probability $p$ is thus symmetric. It is not in general reflexive, nor is it transitive, that is to say, we do not in general have that if

$$P(A \wedge B) = p(A) \cdot p(B) \quad \text{and} \quad p(B \wedge C) = p(B) \cdot p(C)$$

then

$$p(A \wedge C) = p(A) \cdot p(C).$$

To see this just notice that the independence of $A$ and $B$ implies that of $B$ and $A$, but $A$ is not in general independent of itself.

If $A$ and $B$ are independent in a $T$-probability $p$, then so are $A$ and not-$B$, not-$A$ and $B$, and not-$A$ and not-$B$. This is just as we should expect from the intuitive foundation of independence: If $A$ has no effect on $B$, it will have no effect on not-$B$. To see this just notice that

$$p(A \wedge -B) = p(A) - p(A \wedge B)$$

so

$$\frac{p(A \wedge -B)}{p(A)} = 1 - \frac{p(A \wedge B)}{p(B)}$$

and if we assume $A$ and $B$ to be independent, then

$$\frac{p(A \wedge -B)}{p(A)} = 1 - p(B) = p(-B).$$

So, on that assumption, $A$ and not-$B$ are also independent.

An interesting relation between probabilistic independence and logical consistency is this. If the $T$-probability $p$ assigns neither zero nor one to either $A$ or $B$, and if $A$ and $B$ are independent in $p$, then all the constitutions of $A$, $B$ are $T$-consistent. For if $A$ and $B$ are independent in $p$, and $0 < p(A), p(B) < 1$, then

$$\begin{aligned}
p(A \wedge B) &= p(A)\cdot p(B) > 0\\
p(A \wedge -B) &= p(A)\cdot p(-B) > 0\\
p(-A \wedge B) &= p(-A)\cdot p(B) > 0\\
p(-A \wedge -B) &= p(-A)\cdot p(-B) > 0
\end{aligned}$$

and since each of these conjunctions has non-zero probability, each is $T$-consistent.

When we consider sets of sentences of size larger than two, the concept of independence becomes more complicated, because of the increased number of possible dependencies. To see this we formulate two definitions.

Let $\langle \Omega, T, p \rangle$ be a probability system. Then the subset $X$ of $\Omega$ is *weakly independent* in the conjunctive extension of $p$ just in case

$$\hat{p}(X) = \prod_{A \in X} p(A).$$

Let $\langle \Omega, T, p \rangle$ be a probability system. Then the subset $X$ of $\Omega$ is *thoroughly independent* in $\hat{p}$ just in case every subset of $X$ is weakly independent in $\hat{p}$.

That sets of sentences may be weakly but not thoroughly independent may be seen by means of examples, which are not, however, quite trivial to generate. An example is provided in the appendix on measure theory.

The distinction between weak and thorough independence is epistemically important. The epistemic import of the distinction has something to do with this: One may take the set $\{A_1, \ldots, A_k\}$ to be weakly independent, so, his beliefs adhering to the probability $p$,

$$\hat{p}(\{A_1, \ldots, A_k\}) = p(A_1)\cdot p(A_2)\cdot \cdots \cdot p(A_k)$$

and he may also take the set $\{A_1, \ldots, A_{k-1}\}$ to be weakly independent, so that he takes the chance of $A_k$ to be the same as that of $A_k$ given $A_1 \wedge \cdots \wedge A_{k-1}$. But it does not follow in general that he therefore takes the chance of each member of the set to be the same as its chance given the other members, that would require that the set of those other members should be weakly independent in $\hat{p}$. But the thorough independence of a set of sentences means that the chance of any one or several of them is the same in the presence or absence of any or all of the others.

Among the more philosophically interesting consequences of the definition of thorough independence is its characteristic of compactness. We have:

THEOREM 3.1 If $\langle \Omega, T, p \rangle$ is a probability system, then a subset $X$ of $\Omega$ is thoroughly independent in $\hat{p}$ just in case each of its finite subsets is thoroughly independent in $\hat{p}$.

*Proof:* We need only argue that the thorough independence of every finite subset is sufficient for that of $X$.

If $Z$ is any subset of $X$, let $\{Z_i\}$ be a covering nest of $Z$. Then

$$p(Z) = \lim_{i \to \infty} \hat{p}(Z_i).$$

If we assume every finite subset of $X$, hence of $Z$, to be thoroughly independent in $\hat{p}$, then

$$p(Z) = \lim_{i \to \infty} \hat{p}(Z_i) = \lim_{i \to \infty} \prod_{A \in Y_i} p(A).$$

Thus

$$\hat{p}(Z) = \prod_{i=1}^{\infty} p(A_I)$$
$$= \prod_{A \in Z} p(A).$$

This result means that questions of the thorough independence of denumerable infinities of sentences may always be reduced to questions about their finite subsets. It stands in some analogy to the compactness property of inconsistency of ordinary logic and shares with it a reductionistic character. It allows us, for example, to suppose that what is before the mind is always a finite totality, but still to make sense of questions of the thorough independence of an infinite set. A man might be said to hold a set to be thoroughly independent if he held each of its finite subsets to be so: We should thus convert a question about a belief with an infinite object into a question about an infinite number of beliefs each about a finite object. That is, from virtually any point of view, a reduction of complexity, for it involves the elimination of an intentional infinity for an infinity of intentions.

We have also the obvious

COROLLARY: If $\langle \Omega, T, p \rangle$ is a probability system, then a subset $X$ of $\Omega$ is thoroughly independent in $\hat{p}$ just in case every finite subset of $X$ is weakly independent in $\hat{p}$.

This discussion is continued and extended in the next section.

## V.4. Conditional probability

The concepts of probability are in many ways analogous to those of logic. Carnap thought of probabilistic relations as generalizations of the deductive logical relations, and that is to some extent true of the present approach, though in a different sense. In any event, one of the most obvious and profound analogies is that of conditional probability to implication. The definition of conditional probabilities for conjunctive extensions of $T$-probabilities is pretty straight-forward, and, at least from one plausible point of view, is a generalization within which the previously defined measures occur as a special case.

Let $\langle \Omega, T, p \rangle$ be a probability system. Then if $X$ is a subset of $\Omega$ with $\hat{p}(X) \neq 0$, the *conditionalization* of $\hat{p}$ to $X$, written $\hat{p}_X$, is defined for all subsets $Y$ of $\Omega$,

$$\hat{p}_X(Y) = \frac{\hat{p}(X \cup Y)}{\hat{p}(X)}.$$

We have first a theorem in several parts giving some general characteristics of $\hat{p}_X$, in which we see that it behaves as one should expect.

THEOREM 4.1. Let $\langle \Omega, T, p \rangle$ be a probability system, let $X$, $Y$, $Z$ with or without subscripts range over subsets of $\Omega$, and let $A$, $B$ range over members of $\Omega$. Assume that conditionalizations of $p$ are in each case defined. Then

(1) $\quad 0 \leqslant \hat{p}_X(Y) \leqslant 1.$

*Proof:* $\hat{p}$ is a $T$-probability and is thus monotone non-increasing in the subset relation.

(2) If $X \cup \{A, B\}$ is $T$-inconsistent, then

$$\hat{p}_X\{A \vee B\} = \hat{p}_X\{A\} + \hat{p}_X\{B\}.$$

*Proof:* Let $\{X_i\}$ be a covering nest of $X$. Since $\hat{p}(X) \neq 0$, $\hat{p}(X_i) \neq 0$ for each $i$. If $C$ is any sentence in $\Omega$, then, since $A$ and $B$ are assumed to be $T$-incompatible, $C \wedge A$ and $C \wedge B$ are also $T$-incompatible. Hence

$$p(C \wedge (A \vee B)) = p((C \wedge A) \vee (C \wedge B)) = p(C \wedge A) + p(C \wedge B).$$

Thus for $X_i \in \{X_i\}$,

$$p((A \vee B) \wedge \bigwedge X_i) = p(A \wedge \bigwedge X_i) + p(B \wedge \bigwedge X_i).$$

The sequences

$$\Lambda, \{A \vee B\}, \{A \vee B\} \cup X_1, \{A \vee B\} \cup X_2, \ldots.$$
$$\Lambda, \{A\}, \{A\} \cup X_1, \{A\} \cup X_2, \ldots.$$
$$\Lambda, \{B\}, \{B\} \cup X_1, \{B\} \cup X_2, \ldots.$$

are covering nests of $X \cup \{A \vee B\}$, $X \cup \{A\}$, $X \cup \{B\}$, respectively, so

$$\inf \{p((A \vee B) \wedge \bigwedge X_i)\} = \inf \{p(A \wedge \bigwedge X_i)\} + \inf \{p(B \wedge \bigwedge X_i)\}$$

and

$$\hat{p}(X \cup \{A \vee B\}) = \hat{p}(X \cup \{A\}) + \hat{p}(X \cup \{B\})$$

from which (2) follows by the definition of $\hat{p}_X$.

(3) If $X$ $T$-implies $Y$, then $\hat{p}_X(Y) = 1$.

(4) $\hat{p}_X(Y \cup Z) = \hat{p}_{X \cup Y}(Z) \cdot \hat{p}_X(Y)$.

(5) $\hat{p}_\Lambda(X) = \hat{p}(X)$.

We have the finite generalization of (4):

(6) $\hat{p}_X(Y_1 \cup \ldots \cup Y_n) = \hat{p}_X(Y_1) \cdot \hat{p}_{X \cup Y_1}(Y_2) \ldots \hat{p}_{X \cup Y_1 \cdots \cup Y_{n-1}}(Y_n)$

which is a computational consequence of the definition. It is perhaps more perspicuous, and certainly neater, in abbreviated form:

$$\hat{p}_X \bigcup_{i=1}^{n} \{Y_i\} = \prod_{k=1}^{n} \hat{p}_{X \cup \bigcup_{i=0}^{k-1} \{Y_i\}}(Y_k)$$

letting $Y_0 = \Lambda$.

Further, we have the generalization of (6) to the denumerably infinite case, demonstrating the infinite capacity of conjunctive extensions.

(7) Let $Y_1, \ldots$ be a denumerable sequence of subsets of $\Omega$, then

$$\hat{p}_X \bigcup_{i=1}^{\infty} \{Y_i\} = \prod_{k=1}^{\infty} \hat{p}_{X \cup \bigcup_{i=0}^{k}\{Y_i\}}(Y_k)$$

where $Y_0 = \Lambda$.

*Proof:* The proposition will follow from

$$\text{(i)} \qquad \hat{p}[X \cup \bigcup_{i-1}^{\infty} \{Y_i\}] = \lim_{k \to \infty} \hat{p}[X \cup \bigcup_{i=1}^{k} \{Y_i\}].$$

To establish (i), let

$$Z = X \cup \bigcup_{i=1}^{\infty} \{Y_i\},$$

and for each $k$, let

$$Z^k = X \cup \bigcup_{i=1}^{k} \{Y_i\}.$$

So $Z$ is the denumerable union of the sets $Z^k$. For each $k$, $\hat{p}(Z^k) \geqslant \hat{p}(Z)$, and, clearly, by picking $k$ sufficiently large, the value $\hat{p}(Z^k)$ may be brought arbitrarily close to $\hat{p}(Z)$. Indeed, any finite subsets of $Z$ are finite subsets of some $Z^k$, and thus the limit of the values of $\hat{p}$ over finite subsets of $Z$ is approached, as $k$ increases, by the limit of the values of $\hat{p}$ over finite subsets of $Z^k$. The left side of (i) is just the first of these limits, and the quantity

$$\hat{p}(Z^k) = \lim_{j \to \infty} \hat{p}(Z_j^k)$$

where $Z_j^k$ ranges over the finite subsets of $Z^k$, is just the matrix

$$\hat{p}\left[X \cup \bigcup_{i=1}^{k} \{Y_i\}\right]$$

of the right side of (i)

Now the proposition follows from (i) and (6), for

$$\prod_{k=1} \hat{p}_{X \cup \bigcup_{i=0}^{k}\{Y_i\}}(Y_k) = \lim_{n \to \infty} \prod_{k=1}^{n} \hat{p}_{X \cup \bigcup_{i=0}^{k}\{Y_i\}}(Y_k)$$

which by the finite form (6) yields

$$= \lim_{n \to \infty} \hat{p}_X \bigcup_{i=1}^{n} \{Y_i\}$$

$$= \frac{\lim_{n \to \infty} \hat{p}(X \cup \bigcup_{i=1}^{n} Y_i)}{\hat{p}(X)}$$

and by (i)

$$= \frac{\hat{p}[X \cup \bigcup_{i=1}^{\infty} \{Y_i\}]}{\hat{p}(X)}$$

$$= \hat{p}_X \bigcup_{i=1}^{\infty} \{Y_i\}$$

We can define the useful concept of relative independence: If $\langle \Omega, T, p \rangle$ is a probability system, $X$ and $Y$ are subsets of $\Omega$, and $\hat{p}(X) \neq 0$, then $Y$ is *weakly independent* of $X$ in $\hat{p}$, if

$$\hat{p}_X(Y) = \hat{p}(Y)$$

and $Y$ is thoroughly independent of $X$ in $\hat{p}$ if $Y$ is weakly independent of every subset of $X$ in $\hat{p}$.

The expected compactness theorem is forthcoming.

THEOREM 4.2. If $\langle \Omega, T, p \rangle$ is a probability system, $\hat{p}(X) \neq 0$, and the subset $Y$ of $\Omega$ is weakly independent of every finite subset of $X$ in $\hat{p}$, then $Y$ is thoroughly independent of $X$ in $\hat{p}$.

*Proof*. The theorem will follow from a

LEMMA: $\hat{p}_X(Y) = \inf_j \{\hat{p}_{X_j}(Y)\}$ where $X_j$ ranges over the finite subsets of $X$.

This yields the theorem straightway, for on the assumption of the conditions of the theorem; we have that

$$\hat{p}_{X_j}(Y) = \hat{p}(Y)$$

for each finite $X_j \subseteq X$. So – assuming the lemma –

$$\hat{p}_X(Y) = \inf_j \{\hat{p}_{X_j}(Y)\} = \inf_j \{\hat{p}(Y)\} = \hat{p}(Y).$$

Thus we attend to the lemma. We argue first that

$$\text{(i)} \qquad \hat{p}(X \cup Y) = \inf \{\hat{p}(X_j \cup Y)\}.$$

To see this, let $Y_i$ and $Z_k$ range over the finite subsets of $Y$ and $X \cup Y$ respectively. We may assume a one-to-one correspondence between the indices $k$ and pairs $(i, j)$ of indices such that

$$X_j \cup Y_i = Z_k.$$

That is, for each finite subset of $X \cup Y$, there are finite subsets of $X$ and of $Y$ of which it is the union, and conversely. Thus, if we take the infimum over pairs $(i, j)$ of

$$\hat{p}(X_j \cup Y_i)$$

this bound will be the same as the infimum over indices $k$ of

$$\hat{p}(Z_k).$$

The first of these is the greatest lower bound of the set of greatest lower bounds.

$$\{\inf_i \{\hat{p}(X_j \cup Y_i)\}\}$$

that is, it is

$$\inf_j \{\inf_i \{\hat{p}(X_j \cup Y_i)\}\}$$

which is equal to

$$\inf_k \hat{p}(Z_k) = \hat{p}(X \cup Y)$$

and

$$\inf_i \{\hat{p}(X_j \cup Y_i)\} = \hat{p}(X_j \cup Y)$$

for each $j$, so

$$\begin{aligned}\hat{p}(X \cup Y) = \inf_k \{\hat{p}(Z_k)\} &= \inf_j \{\inf_i \{p(X_j \cup Y_i)\}\} \\ &= \inf_j \{\hat{p}(X_j \cup Y)\}\end{aligned}$$

which establishes (i).

Thus

$$\frac{\hat{p}(X \cup Y)}{\hat{p}(X)} = \frac{\inf_j \{\hat{p}(X_j \cup Y)\}}{\inf_j \{\hat{p}(X_j)\}}$$

Both these bounds are defined, and the assumption that $\hat{p}(X) \neq 0$ entails that no $\hat{p}(X_j)$ is zero. Thus the quotient is

$$\inf_j \left\{ \frac{\hat{p}(X_j \cup Y)}{\hat{p}(X_j)} \right\}$$

so,

$$\hat{p}_X(Y) = \inf \{\hat{p}_{X_j}(Y)\}$$

which establishes the theorem.

This theorem has the consequence that whether or not a given set of sentences is independent of another set may be decided by raising the question only with respect to finite subsets of the second set. Thus, on the appropriate assumptions, if no finite evidence would change your belief in a sentence, or a set of them; then no evidence would change that belief. It is worth remarking, I think, that this result depends not at all upon compactness of the base logic, it holds as well if that logic is not compact, and is rather a consequence of defining the infinite extensions in terms of limit concepts.

## V.5. Transparency and Monotonicity

As might be expected from the definition, the conjunctive extension of a $T$-probability $p$ is $T$-transparent if $T$-inconsistency is compact in the field in question. If $X$ and $Y$ are subsets of a field $\Omega$, then $X$ $T$-implies $Y$ just in case $X$ $T$-implies every member, and hence every subset, of $Y$. If $T$ is compact in $\Omega$ and $X$ $T$-implies $Y$ then any finite subset of $Y$ is $T$-implied by some finite subset of $X$. Thus the following:

THEOREM 5.1. Let $\langle \Omega, T, p \rangle$ be a compact probability system. Then if the subset $X$ of $\Omega$ $T$-implies the subset $Y$ of $\Omega$, $\hat{p}(X) \leqslant \hat{p}(Y)$ ($\hat{p}$ is monotone non-decreasing with $T$-implication.)

*Proof:* Assume that $X$ $T$-implies $Y$, and let $\{X_i\}$, $\{Y_j\}$ be covering nests of $X$ and $Y$. If $Y_j$ is any member of $\{Y_j\}$ then some finite subset of $X$ $T$-implies $Y_j$, and thus some $X_i$ in the nest $\{X_i\}$ $T$-implies $Y_j$. The monotonicity of $p$ (III·2·1) entails then that $\hat{p}(X_i) \leqslant \hat{p}(Y_j)$. Thus for each $Y_j \in \{Y_j\}$, there is some $X_i \in \{X_i\}$ such that $\hat{p}(X_i) \leqslant \hat{p}(Y_j)$. Thus

$$\inf \{\hat{p}(X_i)\} \leqslant \inf \{\hat{p}(Y_j)\}$$
$$\hat{p}(X) \leqslant \hat{p}(Y).$$

COROLLARY: Under the same assumptions, if $X$ and $Y$ are $T$-equivalent, then $\hat{p}(X)=\hat{p}(Y)$.

And we may now move to the monotonicity of $\hat{p}$;

THEOREM 5.2. Let $\langle\Omega, T, p\rangle$ be a compact probability system. If $X$ and $Y$ are subsets of $\Omega$ such that no maximal $T$-consistent subset of $\Omega$ is intersected by $X$ which is not also intersected by $Y$, then $\check{p}(X)\leqslant\check{p}(Y)$.

*Proof:* Assume the conditions of the theorem. Let

$$\overline{X} = \{-A \mid A \in X\}$$

and similarly for $\overline{Y}$. Then – since the sets $\Phi$ are maximal and $T$-consistent – for any $\Phi \in M(\Omega)$

$$X \cap \Phi \neq \Lambda \Leftrightarrow \overline{X} \nsubseteq \Phi$$

and thus

$$\overline{X} \nsubseteq \Phi \Rightarrow \overline{Y} \nsubseteq \Phi \quad \text{for each } \Phi$$

so

$$\overline{Y} \subseteq \Phi \Rightarrow \overline{X} \subseteq \Phi \quad \text{for each } \Phi$$

and $\overline{Y}$ $T$-implies $\overline{X}$. Thus, by the previous theorem;

$$\hat{p}(\overline{Y}) \leqslant \hat{p}(\overline{X})$$

from which the theorem follows.

COROLLARY: If $X$ and $Y$ intersect all and only the same maximal $T$-consistent subsets of $\Omega$; then $\check{p}(X)=\check{p}(Y)$.

These results combine with the preceding discussion of independence to yield some consequences about conditional measures. If $\hat{p}_X$ is a conditionalization we sometimes speak of $X$ and $Y$ as antecedent and consequent respectively of $\hat{p}_X(Y)$.

THEOREM 5.3. In (1)–(5) let $\langle\Omega, T, p\rangle$ be a compact probability system. Let $X$, $Y$, and $Z$ range over subsets of $\Omega$, and assume that conditionalizations are defined.

(1) If $(X\cup Y)$ $T$-implies $(X\cup Z)$ then $\hat{p}_X(Y)\leqslant\hat{p}_X(Z)$.

*Proof:* From the monotonicity of $\hat{p}$ and the definition of the conditionalization.

(2) If $(X\cup Y)$ is $T$-equivalent to $(X\cup Z)$ then $\hat{p}_X(Y)=\hat{p}_X(Z)$.

(3) If $Y$ $T$-implies $Z$ then $\hat{p}_X(Y) \leqslant \hat{p}_X(Z)$.

(4) If $Y$ and $Z$ are $T$-equivalent then $\hat{p}_X(Y) = \hat{p}_X(Z)$.

Conditionalizations are also transparent at the antecedent whenever defined:

(5) If $X$ and $Y$ are $T$-equivalent then $\hat{p}_X(Z)$.

But they are not in general monotonic in either direction at the antecedent. It is easy to see the reason for this in logical terms by considering a simple case. The null set serves as a tautology for conditionalizations. Thus if $A$ is any sentence $\hat{p}(\{A\}) = \hat{p}_\Lambda(\{A\})$. And the null set is $T$-implied by every sentence. Thus monotonicity of $\hat{p}_X$ at the antecedent would entail an inequality between $\hat{p}(\{A\})$ and $\hat{p}_{\{B\}}(\{A\})$. That is to say, between

$$p(A \wedge B) + p(A \wedge -B)$$

and

$$\frac{p(A \wedge B)}{p(A \wedge B) + p(-A \wedge B)}$$

But there are examples of probabilities in which these are equal

| | $A \wedge B$ | $A \wedge -B$ | $-A \wedge B$ | $-A \wedge -B$ |
|---|---|---|---|---|
| $p_1$ | $\frac{1}{2}$ | 0 | $\frac{1}{2}$ | 0 |

$$p_1(A) = \tfrac{1}{2} = p_{1_B}(A) = \frac{p_1(A \wedge B)}{P_1(A \wedge \mathrm{B}) + p_1(-A \wedge B)}$$

in which the first exceeds the second:

| | $A \wedge B$ | $A \wedge -B$ | $-A \wedge B$ | $-A \wedge -B$ |
|---|---|---|---|---|
| $p_2$ | 0 | $\frac{1}{2}$ | $\frac{1}{2}$ | 0 |

$$p_2(A) = \tfrac{1}{2} > p_{2_B}(A) = 0$$

and in which the second exceeds the first:

| | $A \wedge B$ | $A \wedge -B$ | $-A \wedge B$ | $-A \wedge -B$ |
|---|---|---|---|---|
| $p_3$ | $\frac{1}{2}$ | 0 | $\frac{1}{4}$ | $\frac{1}{4}$ |

$$p_3(A) = \tfrac{1}{2} < p_{3_B}(A) = \tfrac{2}{3}.$$

If we represent probabilities by areas in a diagram

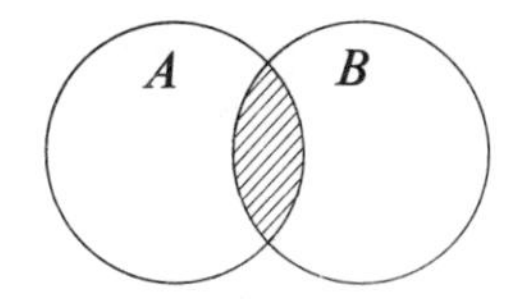


then the conditional probability of $A$ given $B$ is the ratio of the shaded area to the area $B$, and this quotient can be varied independently of the size of $A$.

Epistemically, the reason for the lack of monotonicity of any simple sort at the antecedent is not hard to state and is profound in its consequences: One proposition may provide more evidence for another than does a consequence of it, and less evidence than does another consequence of it.

## V.6. Systems with finite bases; indifference

In this section some theorems are developed for the important special case of fields with finite $T$-bases. Of particular interest are *indifference functions* on such fields which maximize independence characteristics in interesting ways relevant to the applicability of constraints on changes in belief. This development begins by recalling and expanding the earlier characterization of bases.

If $X$ is a finite $T$-base for a field $\Omega$, then if $A$ is any $T$-consistent sentence in $\Omega$, and $C_A$ the set of those constitutions of $X$ each of which $T$-implies $A$, then $A$ is $T$-equivalent to the disjunction[2]

$$\bigvee_{X_i \in C_A} \{\wedge X_i\}.$$

Since the number of distinct constitutions of a finite set $X$ is finite, the number of distinct disjunctions of them is also finite, and thus the sentences of a field with a finite $T$-base are partitioned into a finite number of $T$-equivalence classes; one class including each of the disjunctions of conjunctions of constitutions of the $T$-base, and one class including just the $T$-inconsistent sentences.

It was noted in chapter III, in the theorem mentioned above, that a field with a finite $T$-base has only finitely many maximal $T$-consistent subsets, and that these subsets correspond in a one-to-one manner with constitutions of the $T$-base, to which they are respectively $T$-equivalent. This permits the partitioning of the subsets of $\Omega$ into $T$-equivalence classes as follows: Let $Y$ be any $T$-consistent subset of $\Omega$, and let $\Phi_1, \dots, \Phi_k$ be all the maximal $T$-consistent extensions of $Y$. Let $X_1, \dots, X_k$ be the constitutions of the $T$ base $X$ to which $\Phi_1, \dots, \Phi_k$ are respectively $T$-equivalent, and the existence of which is guaranteed by Theorem III.1.4. Then define

$$C_Y = \{X_1, \dots, X_k\}.$$

Now each $X_i \in C_Y$ $T$-implies $Y$: $\wedge X_i$ $T$-implies every $A$ in $C_Y$, for each $i$ from 1 to $k$. Thus the disjunction of these conjunctions $T$-implies every $A \in Y$, and

$$\bigvee_{X_i \in C_Y} \{\wedge X_i\} = \wedge X_i \vee \dots \vee \wedge X_k$$

$T$-implies $Y$.

Further, $Y$ $T$-implies this disjunction; any maximal $T$-consistent extension of $Y$ is one of the $\Phi_i$, and thus some $X_i \in C_Y$ is s subset of this $\Phi_i$. Thus

$$\wedge X_1 \vee \dots \vee \wedge X_k$$

is a member of every maximal $T$-consistent extension of $Y$, and is hence $T$-implied by $Y$. If $Z$ is any subclass of $\Omega$ which is $T$-equivalent to $Y$, then $Z$ will also be $T$-equivalent to this disjunction. Thus we have a

COROLLARY TO THEOREM III.1.4 If $X$ is a finite $T$-base for a field $\Omega$ and $Y$ is any $T$-consistent subset of $\Omega$, then there is a set $C_Y$ of constitutions of $X$ such that $Y$ is $T$-equivalent to the sentence

$$\bigvee_{X_i \in C_Y} \{\wedge X_i\}.$$

This corollary opens the way for a useful theorem about probabilities. Before developing it, there is a further character of fields with finite $T$-bases to be remarked: The transparency results of preceding sections depend upon the compactness of the base logic. The development will proceed more smoothly, and require less explicit assumptions, if we see first that compactness of $T$-inconsistency is entailed by the presence of a finite $T$-base. We do this in a

LEMMA A: If the denumerable field $\Omega$ has a finite $T$-base, then $T$-inconsistency is compact in $\Omega$.

This lemma is not as uninteresting as it may at first appear. It may diminish in apparent triviality upon remarking that whether $\Omega$ has a finite $T$-base or not, every $T$-inconsistent subset of $\Omega$ is $T$-equivalent to a sentence of $\Omega$, but this sentence may not itself be a member of the inconsistent set in question.

The proof is gentle, but not indelicate, and is given here in some detail. It will depend upon constructing a sequence $f_0 \supseteq f_1 \supseteq \ldots$ of sets of maximal $T$-consistent subsets of $\Omega$. Each $f_i$ is a set, the members of which are maximal $T$-consistent subsets of $\Omega$, and the sequence $\{f_i\}$ is non-ascending in the sense that each $f_i$ is a subset of all its predecessors. In general, it is quite possible that each set $f_i$ be non-null, but that the intersection $\bigcap_i \{f_i\}$, of all the sets in the sequence be null. That is not paradoxical, and in a later section an example of such a sequence is provided.

To see how sequences of this sort relate to compactness, let $X$ be a set of sentences, and $\{X_i\}$ a covering nest of $X$. For each $i$; let $f_i$ be the set of maximal $T$-consistent extensions of $X_i$, and let $f$ be the set of maximal $T$-consistent extensions of $X$. Then, since $\{X_i\}$ is a nest, and $X_0 \subseteq X_1 \subseteq \ldots$ $f_0 \supseteq f_1 \supseteq \ldots$ and the sequence $\{f_i\}$ is just as described above. We have, letting $\Phi$ range over maximal $T$-consistent subsets of $\Omega$.

(i) If $\Phi \in \cap_i \{f_i\}$, then every finite subset of $X$ is a subset of $\Phi$, and $X \subseteq \Phi$, so $\Phi \in f$. Thus $\bigcap_i \{f_i\} \subseteq f$.

(ii) If $\Phi \in f$ then $\Phi$ is a maximal $T$-consistent extension of $X$, hence of every $X_i$, so $\Phi \in \bigcap_i \{f_i\}$. Thus $f \subseteq \bigcap_i \{f_i\}$ and $f = \bigcap_i \{f_i\}$.

If $T$-inconsistency is *not* compact in $\Omega$, then there will be some $T$-inconsistent $X \subseteq \Omega$, every finite subset of which is $T$ consistent. That is to say, that in the sequence $\{f_i\}$ associated with the covering nest $\{X_i\}$ of $X$, each set $f_i$ of maximal $T$-consistent subsets will be non-null, but the intersection $f$, of all these sets, the set of maximal $T$-consistent extensions of $X$, will be null.

If, on the other hand, $T$ inconsistency *is* compact in $\Omega$ then if $f = \Lambda$, some $f_i$ must be null as well.

Thus, assuming the appropriate correspondences among $X$, $\{X_i\}$, $f$, and $\{f_i\}$;

(a) $T$-inconsistency is compact in $\Omega$ if and only if, for every subset $X$ of $\Omega$, if $f = \Lambda$, then for some $i$, $f_i = \Lambda$.

And the lemma will follow from

(b) If $\Omega$ has a finite $T$-base, then for every subset $X$ of $\Omega$, if no $f_i$ is null, then $\bigcap_i \{f_i\} \neq \Lambda$.

To see that (b) is true, let $n_0, \ldots, n_i, \ldots$ and $n$ be the sizes of $f_0$, $f_1, \ldots,$

$f_i, \ldots$ and $f$ respectively. $n_i$ is just the number of maximal $T$-consistent sets in $f_i$. It is a consequence of theorem III.1.5 that, on the assumption that $\Omega$ has a finite $T$-base, the number of maximal $T$-consistent subsets is finite, and thus that every $f_i$ is finite. Also, in view of the construction of the sequence $\{f_i\}$, $n_0 \geqslant n_1 \geqslant \ldots \geqslant n$. Since $n_0$ is finite, the numbers $n_i$, $n_{i+1}, \ldots$ cannot keep decreasing indefinitely, indeed, the number of cases in which $n_{i+1}$ is less than $n_i$ can be no greater than $n_0$. Thus

(c) If $\Omega$ has a finite $T$-base, then for some $i_0$, for all $i > i_0$, $n_i = n_{i_0}$.

Now assume that $\Omega$ has a finite $T$-base, let $X$ be a subset of $\Omega$ and let $\{X_i\}$, $\{f_i\}$, $f$, and the numbers $\{n_i\}$ be as described above. To establish (b) assume that no $f_i$ is null. Then, for all $i$, $n_i > 0$. And by (c), for some $i_0$, for all $i \geqslant i_0$, $f_i = f_{i_0} \neq \Lambda$. $f_{i_0}$ is a subset of all earlier members of the sequence $\{f_i\}$. Thus

$$\Lambda \neq f_{i_0} \subseteq \bigcap_i \{f_i\}$$

so $f = \bigcap_i \{f_i\}$ is not null. This establishes (b). Now the lemma follows from (a) and (b).

This lemma, in concert with theorems 5.1 and 5.2 of the previous section and their corollaries, supports analogues to those results in which the assumption of compactness is replaced by that of finitude of the base.

THEOREM 6.1: If $\langle \Omega, T, p \rangle$ is a probability system with a finite $T$-base, then $\hat{p}$ is monotone non-increasing for $T$-implication.

COROLLARY: Under the same assumptions, $\hat{p}$ is transparent for $T$-equivalence.

THEOREM 6.2. If $\langle \Omega, T, p \rangle$ is a probability system with a finite $T$-base, then if $X$ and $Y$ are subsets of $\Omega$ such that $X$ intersects no maximal $T$-consistent subset of $\Omega$ which is not also intersected by $Y$, then $\check{p}(X) \leqslant \check{p}(Y)$.

COROLLARY: Under the same assumption, if $X$ and $Y$ intersect just the same maximal $T$-consistent subsets of $\Omega$ then $\check{p}(X) = \check{p}(Y)$.

It may also be remarked that the lemma permits employment of theorem 5.3 on conditional probabilities.

We can now state the fundamental theorem on systems with finite $T$-bases.

THEOREM 6.3. Let $\langle \Omega, T, p \rangle$ be a probability system with a finite $T$-base $X$. Then if $Y$ is any $T$-consistent subset of $\Omega$ and $C_Y$ is the set of those constitutions of $X$ each of which $T$-implies $Y$,

$$\text{(i)} \qquad \hat{p}(Y) = p\Big(\bigvee_{X_i \in C_Y} \{\Lambda X_i\}\Big)$$

$$\text{(ii)} \qquad = \sum_{X_i \in C_Y} \hat{p}(X_i)$$

$$\text{(iii)} \qquad = \check{p}\{\wedge X_i | X_i \in C_Y\}.$$

*Proof:* By the corollary to theorem 6.1, $\hat{p}$ is transparent for $T$-equivalence. Thus (i) follows from the preceding corollary.

(ii): The members of $C_Y$ are pairwise $T$-incompatible, so the right side of (i) is equal to that of (ii) by the additivity of $p$ and the definition of $\hat{p}$.

(iii): The right sides of (i) and (iii) are equal by the definition of $\check{p}$.

One of the central laws of the classical theory of probability is the principle of indifference. This principle is not easy to state correctly. It is usually formulated in terms of some other principle: Leibniz and La Place, for example, put in roughly as follows: If there is not sufficient reason for either of two alternatives to be more likely than the other, then they are equally likely. Hume's formulation depends upon understanding the composition of chances:

> Since, therefore, an entire indifference is essential to chance, no one chance can possibly be superior to another, otherwise than as it is composed of a superior number of equal chances.[3]

Several different sorts of difficulties have been found with the principle. The first sort consists of objections on the grounds that the principle is unclear or that it may, because of this, lead to contradictory probability assignments: Thus the Leibniz-LaPlace form of the principle depends upon some clarification of sufficiency of reason which would provide an appropriate equivalence relation, not resulting in violation of the transitivity of equiprobability, for example. And it has been argued against the principle that by following it one may conclude both that the chance of tossing two heads in a row is one fourth (if the equiprobable alternatives are thought of as the four possible two-toss sequences) and that the chance is one third (if the alternatives are thought of as zero, one, or two heads).

Problems of this latter sort may or may not be fundamental to the principle of indifference. They may as well be taken as difficulties with another principle to which the classical theorists held, in general without much criticism, namely that the probabilities of events are uniquely determined. That each event has a probability which is given as a number between zero and one, and that this number gives the chance or probability that the event will occur. La Place held that probability was relative to knowledge and ignorance, so that relative to different bodies of knowledge one could consistently have different probabilities, but he took the notions of *possible* and *favorable* case to be determined, by sufficiency or insufficiency of reason, given a body of knowledge, and he assumed that the ratio of cases favorable to those possible was uniquely determined in turn.

It is clear, I think, that the assumption that probability is uniquely determined may be separated from the principle of indifference. It is also clear how that assumption could reinforce the principle, if not lead to it: If probability is assumed to be uniquely determined, then principles will be sought which will narrow the class of possible probability assignments. Something like this seems admittedly and explicitly to have motivated Carnap's early work in probability.[4] He thought of conditional probability, in one of its forms, as a quantitative logical relation which holds between sentences; the degree to which one sentence confirms the other. And he took the task of the logical theory of probability to be the articulation of the principles, among which he at one time included a form of the principle of indifference, which would properly determine this degree for given pairs of sentences.

Contemporary writers, including Carnap in his later work, usually reject the assumption that probability is uniquely determined. Subjectivists, in particular on the basis of the equivalence of coherence and the definition of normal measures on the appropriate Tarski-Lindenbaum Algebra, have frequently argued that the determination of probability can go no further than this, augmented, perhaps, by the requirement that only impossible propositions be assigned zero (in view of the equivalence of this last constraint to the possibility of positive gain). These arguments have even gone to the point of rejecting the restriction of denumerable additivity.[5]

In any event, the fact that the principle of indifference in company with other plausible principles, led to distinct probability ascriptions, is a shortcoming of that principle only if one is looking for a definition of *the* probability of events, and if the principle of indifference is viewed rather as a necessary condition, perhaps not always applicable, then this ceases to be an objection against it.

Another difficulty with the principle of indifference is that in most traditional formulations it is not easily applicable when there are infinitely many alternatives. Thus Hume's requirement, which refers to the *number* of chances, looks to have the consequence that probability is not defined for events composed of infinitely many chances, or that all such events have the same probability. These are both consequences which it would be wise to avoid.

The principle of indifference can be put, not as a requirement, but as a definition of a class of measures which then turn out to relate to each other in nice ways, and which may also have something to do with induction and changes in belief.

If $\langle \Omega, T, p \rangle$ is a probability system with a finite base $X$, then the constitutions of the base $X$ are, in terms of the logic $T$, logically the strongest sets in the domain of the conjunctive extension of $p$. None of them is $T$-implied by any other $T$-consistent subset of $\Omega$. The basic sentences, the individual members of the constitutions of $X$, have also an important logical property: Those basic sentences which make up any constitution of $X$ form a logically independent set, and this property of logical independence is lost when any other member of $\Omega$ is added to this set. If $X_i$ is a constitution of the base $X$, and if $A$ is any sentence of $\Omega$ not in $X_i$, then some constitution of $X_i \cup \{A\}$ is $T$-inconsistent. The basic sentences of a base $X$ form, in this sense, a maximal logically independent set of sentences, and intuition correctly suggests that $T$-probabilities in which the base is thoroughly independent in the probabilistic sense will have a pleasing and useful symmetry. The proper definition is easy to come by:

Let $\langle \Omega, T, p \rangle$ be a probability system with a finite $T$-base $X$ of size $k$. Then $p$ is a *total indifference probability* on $\Omega$ just in case for each constitution $X_i$ of the $T$-base $X$

$$p(\wedge X_i) = 1/2^k.$$

This definition, as far as it goes, is in the spirit of Hume's principle quoted above. One plausible interpretation of the *composition of an alternative by chances*, if alternatives are thought of as sentences, is the number of consistent and logically strongest sentences which entail the given alternative, so that the number of chances of which an alternative is composed is the number of distinct disjuncts in the disjunction of conjunctions of constitutions of the base to which it is $T$-equivalent. And one alternative will be less, equally, or more likely than another in the total indifference probability accordingly as it is composed, in this sense, of fewer, the same number of, or more chances. For as the preceding theorem shows, the probability in the total indifference measure of any sentence or set of them is the ratio of the number of the constitutions of the base by each of which it is $T$-implied to $2^k$.

A few theorems may help to develop some characteristics of total indifference measures.

THEOREM 6.4. If $\Omega$ is a denumerable field of sentences, $T$ is at least tautological and absolutely consistent, and $\Omega$ has a finite $T$-base, then the total indifference $T$-probability on $\Omega$ is well defined and unique. That is to say, there is just one probability system $\langle \Omega, T, p \rangle$ such that $p$ is a total indifference probability on $\Omega$.

*Proof:* First, in view of the preceding remarks, it may easily be seen that the constraint of the definition determines a $T$-measure on $\Omega$, and, thus, determines also the extension of $p$ to subsets of $\Omega$. For if $A \in \Omega$, then the class $C_A$ of constitutions of the finite $T$-base $X$ each of which $T$-implies $A$ is unique, and thus the value of $p$ for $A$ is uniquely determined. This value is always bounded by zero and one. If $A$ is $T$-necessary, then each of the $2^k$ $T$-consistent constitutions of $X$ $T$-implies $A$, and hence $p(A)=1$. If $A$ and $B$ are $T$-inconsistent, then the classes $C_A$ and $C_B$ are disjoint, and the class $C_{(A \vee B)} = C_A \cup C_B$. Thus the condition of the definition determines a $T$-probability on $\Omega$.

Further, if there are alternative finite $T$-bases for $\Omega$, the total indifference measure is the same, no matter which base is used to define it. This is a consequence of Theorem III.1.4, mentioned at the end of III.1; that theorem entails that if $X$ and $Y$ are both finite $T$-bases for $\Omega$, then to each constitution of $X$ there corresponds a unique $T$-equivalent constitution of $Y$. To see that the probabilities will be the same, let $p$ and $q$ be total indifference probabilities which are uniform over the constitutions of $T$-bases $X$ and $Y$ respectively, and let $A \in \Omega$. Then $p(A)=n/2^k$, where there are just $2^k$ constitutions of $X$, and just $n$ of them $T$-imply $A$. By the correspondence of the constitutions of $Y$ to those of $X$, there are just $n$ constitutions of $Y$ which $T$-imply $A$, and there are just $2^k$ constitutions of $Y$. Thus

$$q(A) = n/2^k = p(A)$$

and $p$ and $q$ agree for all members of $\Omega$.

As a consequence of this theorem, the conjunctive and disjunctive extensions of the total indifference probability on a field with a finite $T$-base are also unique and well defined.

A second important characteristic of total indifference probabilities is given in the following theorem.

THEOREM 6.5: If $\langle \Omega, T, p \rangle$ is a probability system with a finite $T$-base $X$, and $p$ is the total indifference probability, then every constitution of every finite $T$-base for $\Omega$ is thoroughly independent in $p$, and every basic sentence is assigned $\frac{1}{2}$ by $p$.

*Proof:* Let $X$ be a finite $T$-base for $\Omega$ and let the basic sentence $B$ be an element of $X$. Let the size of $X$ be $k$, so there are just $2^k$ constitutions of $X$, and $B$ is a member of just half of these, i.e., of $2^{k-1}$. $B$ is $T$-implied by just those constitutions of $X$ of which it is a member, so – by Theorem 6.4 –

$$p(B) = \sum_{X_i \ T\text{-implies } B} p(\wedge X_i) = (2^{k-1}) \cdot (1/2^k) = \tfrac{1}{2}.$$

Now to establish that every constitution of every $T$-base is thoroughly independent in $p$, it will suffice to show that $X$ is thoroughly independent in $p$, for every constitution of every $T$-base is a $T$-base, and the choice of the $T$-base $X$ was not otherwise constrained. So let $B \in X$. The conjunction of all the members of $X$ save $B$, $\wedge(X - \{B\})$, is $T$-implied by just two constitutions of $X$, and thus

$$p(X - \{B\}) = 2 \cdot (1/2^k) = 1/2^{k-1}$$

so,

$$\hat{p}(X) = \hat{p}(X - \{B\}) \cdot p(B).$$

It follows by induction on the size $k$ of the finite base for $\Omega$ that every $T$-base, and hence every constitution of every $T$-base, is thoroughly independent in $p$.

The definition of total indifference probabilities is a constraint of the same form as those of finite and denumerable additivity. The conditions of the definition do not involve essentially probabilistic concepts. In the case of finite additivity the condition is that $A$ and $B$ should be $T$-incompatible sentences. In the case of the total indifference probability, on a field with a finite $T$-base, the condition is that $X_i$ should be a constitution of the base. In both cases constraints are put on $T$-probabilities, consequent upon conditions which are non-probabilistic. Constraints of this form have a certain generality and a synthetic character which are lacking from definitions or theorems the conditions of which involve probabilistic concepts, the latter being essentially analytic of the concepts involved.

Constraints with non-probabilistic conditions are of philosophical interest because they relate concepts of essentially different sorts, and because, largely in virtue of this, they raise interesting normative questions of justification. If, for example, probability is taken to be the logic of partial belief, then one can ask why such a logic should conform to additivity constraints. Similarly, one can ask why, or in what situations, the logic of partial belief should conform to constraints of total indifference. The first of these questions has as yet apparently found no clear and definitive answer, and one becomes involved, as the discussion in earlier chapters exemplifies, if it does not conclude, in a tangle of presuppositions and issues of principle in the relation of prescription and description. The second question is just as difficult and involved, and even a casual look at the arguments in, say, Hume, Ramsey and Carnap, will convince one of this. Without pretending to advance these discussions, we may perhaps remark upon some characteristics of total indifference probabilities.

First, the option is not always open to adopt or employ such a measure. It is fairly clear that Hume, for example, in the sentence quoted above and in the surrounding discussion, has in mind what was called above a field with a finite base, which is to say that every proposition before the mind may be conceived as a finite disjunction of incompatible alternatives, which alternatives themselves break down no further in this way. That assumption is essentially satisfied in a field with a finite base, but there are also fields, not apparently pathological, and which may plausibly be assumed to give the logic of the beliefs of some reasonable man, which do not have this character. Any field in which there are infinitely many logically independent sentences provides an example.

Two remarks here: First, it need not be assumed that we contemplate an actual infinity in order for this difficulty to arise. We may take the number of independent sentences to be finite, yet be incapable of putting a bound on it. Since the description of a field with a finite base requires exact stipulation of this number, or, at least; of a finite upper bound for it, in such a case the propositions before the mind will not have the appropriate structure, and the injunction to total indifference probabilities will be inapplicable. Second, we may be able to work out an extension of the principle of indifference to infinite cases. Bernoulli measures, in which the field is considered as a limit of fields with finite bases, accomplish this,[6] and by their means we may come to think of partial belief as transparent for the taking of limits, belief in the logical limit of a set of propositions being treated as the limit of beliefs in propositions in the set. (Something like that is behind the definitions of $\hat{p}$ and $\check{p}$.) But that is another problem, and does not affect the inapplicability of the principle as formulated for finite application. This seems a good argument against taking the principle of indifference to be definitive of probability.

In addition to the difficulties of limited application, there is the general problem of the status of the principle: Does it purport to be a description of the actual logic, of the psychology, in the classical sense, or the phenomenology, of belief? Hume's formulation and supporting argument make it seem so, and within that theory of the mind it apparently is so, for Hume's account of partial belief in terms of shared mental force is both natural and powerful in that context. Nor is that account easy to improve upon, though more recent axiomatizations of comparative and interval probabilities look to be promising options, sufficiently compatible with Hume's general assumptions but independent of the principle of indifference. In any event, however, the limits of Hume's theory of the mind need not constrain other accounts, and we have available now theories of belief which, though they do presuppose the additivity condition, do not involve the principle of indifference in a similar way. Thus we should not attribute to it the *a priori* character which the additivity principle seems to have.

## V.7. Probability and Quantifiers

It may be plausibly argued that the quantifiers may serve an essentially universal role in judgment. That is to say, that quantified propositions are not in general equivalent to conjunctions or disjunctions of their instances, even when we pretend to conceive of those conjunctions and disjunctions as infinite in length. Assuming this argument to be correct, so that the interpretation of the quantifiers in terms of their instances does not give an adequate account of their role in the logic of judgment, this does not mean that quantifiers may never serve such a function. Indeed, there is a sense in which a man who believes that something is *S* believes of something that it is *S*. Quine in *Word and Object*[7] has distinguished this sort of belief in terms of its logical properties. The view which I earlier attributed to Hume is that this is the only way in which belief is to be understood, and Russell, in the fourth lecture on logical atomism,[8] seemed to think that belief in quantified propositions was in general to be analyzed in this way, that is to say as not really containing a proposition as a constituent. Ramsey argued against this view in "General propositions and causality".

The simple form of the view, that believing something to be $S$ is just believing of something that it is $S$, will not endure straightforward importation into an account of partial belief: Believing to degree $r$ that something is $S$ cannot be believing of something to degree $r$ that it is $S$. The chance of someone catching the disease is 9/10, but there is no one who has a chance of 9/10 of catching the disease. The device of defining conjunctive and disjunctive belief functions on classes of propositions, however, invites us to describe beliefs in quantified propositions as disjunctive or conjunctive belief in classes of instances of those propositions. It is quite natural, when the import of the quantifiers is assumed to be exhausted in their instances, to think of belief in $\forall x\, S(x)$ and in $\exists x\, S(x)$ as conjunctive and disjunctive belief respectively in the class $\{S(a_i)\}$ of their instances. The definitions and theorems which do this are, with one small but significant obstacle, straight-forward, and require only some minor prior arrangements.

We recall that a logic is *predicative* if it includes ordinary predicate logic, and that if the system $\langle \Omega, T, p \rangle$ is predicative, then the $T$-necessary members of $\Omega$ form a first order theory in the ordinary sense. If $a$ is an individual constant occurring in the sentences of $\Omega$, then the *instantiation* (with respect to $x$) of $\forall x\, A(x)$ or $\exists x\, A(x)$ by $a$ is $A(a)$; the result of replacing the free $x$'s in $A(x)$ by $a$'s.

If $\langle \Omega, T, p \rangle$ is predicative, and $D$ is the class (always assumed to be non-null) of individual constants occurring in the sentences of $\Omega$, then $T$ is *individual complete* in $\Omega$ just in case $\forall x\, A(x)$ is a member of every maximal $T$-consistent subset of $\Omega$ which includes all instantiations of $\forall x\, A(x)$ by constants in $D$. Thus, if $T$ is individual complete in $\Omega$, and $\Phi$ is any maximal $T$-consistent subset of $\Omega$, then

$$\forall x\, A(x) \in \Phi \Leftrightarrow \{A(a) \mid a \in D\} \subseteq \Phi$$

so, if $T$ is individual complete in $\Omega$, then $\forall x\, A(x)$ is $T$-equivalent to $\{A(a) \mid a \in D\}$, and, as a consequence of this, for each $\Phi$,

$$\exists x\, A(x) \in \Phi \Leftrightarrow \{A(a) \mid a \in D\} \cap \Phi \neq \Lambda.$$

The assumption of individual completeness can complicate logical matters, since it involves the loss of compactness. If $T$ is individual complete in $\Omega$, then $\Omega$ may have a $T$-inconsistent subset, every finite subset of which is $T$-consistent. In particular, if $A$ is a one-place predicate, and if the set $X$ is defined

$$X = \{A(a) | a \in D\} \cup \{-\forall x\, A(x)\}$$

then $X$ is $T$-inconsistent, but each of its finite subsets is $T$-consistent. This provides the example promised in the proof of Lemma A of the preceding section: Let $X$ be enumerated

$$X = \{-\forall x\, A(x), A(a_1), A(a_2), \ldots\}$$

where

$$D = \{a_1, a_2, \ldots\}$$

and construct the covering nest $X_i$,

$$\begin{aligned} X_0 &= \Lambda \\ X_1 &= \{-\forall x\, A(x)\} \\ X_2 &= \{-\forall x\, A(x), A(a_1)\} \\ &\vdots \end{aligned}$$

then, where the sequence $\{f_i\}$ is defined as in the support for Lemma A, we have that

(i) for each $i$, $f_i \neq \Lambda$.
(ii) $\bigcap_i \{f_i\} = f = \Lambda$.

As far as probability is concerned, this failure of compactness entails the loss of the transparency property of conjunctive and disjunctive extensions. That is to say, we may have a system $\langle \Omega, T, p \rangle$, in which the measure $p$ on sentences in $\Omega$ is $T$-transparent (since it is a $T$-probability), but in which the extensions of $p$ are not $T$-transparent.

The above example can be improved to show this as well. Define the predicative system $\langle \Omega, T, p \rangle$ as follows: The sentences of $\Omega$ are composed, by quantifications and finite truth-functions, from the single one place predicate $A$, and the denumerable collection $D = \{a_1, a_2, \ldots\}$ of individual constants. $T$ is first-order logic, so $\langle \Omega, T, p \rangle$ is predicative, and is also augmented by the requirement of individual completeness. Thus the set

$$X = \{-\forall x\, A(x)\} \cup \{A(a_i) | a_i \in D\}$$

is $T$-inconsistent.

The $T$-measure $p$ on sentences of $\Omega$ is defined as follows:

For each $n$ define $p$ on the conjunction of the constitutions of $\{A(a_1), \ldots, A(a_n)\}$

$$p(A(a_1) \wedge A(a_2) \wedge \cdots \wedge A(a_n)) = \frac{2^n + 1}{2^{n+1}}$$

and if $\mathfrak{K}_i^n$ is any other constitution of $\{A(a_1), \ldots, A(a_n)\}$, i.e., if any $A(a_j)$ occurs negated in $\mathfrak{K}_i^n$, then

$$p(\bigwedge \mathfrak{K}_i^n) = \frac{1}{2^{n+1}}.$$

Define $p$ on quantified atomic sentences

$$\begin{aligned} p(\forall x\, A(x)) &= p(\forall x\, -A(x)) = 0 \\ p(\exists x\, A(x)) &= p(\exists x\, -A(x)) = 1. \end{aligned}$$

Thus – writing simply '$A_1$' for '$A(a_1)$', etc.

$$\begin{aligned} & p(A_1 \wedge A_2 \wedge A_3 \wedge \forall x\, A(x) \wedge \exists x\, A(x)) = 0 \\ & p(A_1 \wedge A_2 \wedge A_3 \wedge -\forall x\, A(x) \wedge \exists x\, A(x)) = \tfrac{9}{16} \\ & p(A_1 \wedge A_2 \wedge -A_3 -\forall x\, A(x) \wedge \exists x\, A(x)) = \tfrac{1}{16} \\ & p(A_1 \wedge -A_2 \wedge A_3 \wedge -\forall x\, A(x) \wedge \exists x\, A(x)) = \tfrac{1}{16} \\ & p(A_1 \wedge -A_2 \wedge -A_3 \wedge -\forall x\, A(x) \wedge \exists x\, A(x)) = \tfrac{1}{16} \\ & p(-A_1 \wedge A_2 \wedge A_3 \wedge -\forall x\, A(x) \wedge \exists x\, A(x)) = \tfrac{1}{16} \\ & p(-A_1 \wedge A_2 \wedge -A_3 \wedge -\forall x\, A(x) \wedge \exists x\, A(x)) = \tfrac{1}{16} \\ & p(-A_1 \wedge -A_2 \wedge A_3 \wedge -\forall x\, A(x) \wedge \exists x\, A(x)) = \tfrac{1}{16} \\ & p(-A_1 \wedge -A_2 \wedge -A_3 \wedge -\forall x\, A(x) \wedge \exists x\, A(x)) = \tfrac{1}{16} \\ & p(-A_1 \wedge -A_2 \wedge -A_3 \wedge -\forall x\, A(x) \wedge -\exists x\, A(x)) = 0. \end{aligned}$$

In fact, if consideration is restricted to that subfield of $\Omega$, call it $\Omega_3$, which includes just those sentences which include no individual constants other than $a_1$, $a_2$, or $a_3$, these assignments suffice to determine $p$ everywhere on $\Omega_3$. This is a consequence of the equivalence in predicate logic (the proof for the general case is not easy, but a little computation with instances makes this quite plausible) of any first-order consistent sentence of $\Omega_3$ to a disjunction of some or all of the above ten sentences. Thus, for example, for each $a_i$ (continuing to write $A_i$ rather than $A(a_i)$)

$$\begin{aligned} p(\forall x\, A(x) \rightarrow A_i) &= 1 \\ p(\forall x\, (A_i \rightarrow A(x)) &= p(A_i \rightarrow \forall x\, A(x)) \\ &= 1 - p(A_i \wedge -\forall x\, A(x)) \\ &= 1 - p(A_i) \\ &= \tfrac{1}{4} \end{aligned}$$

and for each $n > 0$, in general,

$$p(-\forall x\, A(x) \wedge A_1 \wedge A_2 \wedge \cdots \wedge A_n) = p(A_1 \wedge A_2 \wedge \cdots \wedge A_n) = \frac{2^n + 1}{2^{n+1}}.$$

Of course, our interest in this example is in what happens as $n$ in these formulas increases without bound. Now the constraints of individual completeness, which were inoperative in the finite cases, begin to be felt, and the importance of the logical character of the example becomes striking. In particular, turning to the conjunctive extension of $p$,

$$\begin{aligned} \hat{p}(X) &= \inf\{p(\bigwedge X_i)\} = \inf\{p(-\forall x\, A(x) \wedge A_1 \wedge \cdots \wedge A_i)\} \\ &= \inf\{1, \tfrac{3}{4}, \tfrac{5}{8}, \tfrac{9}{16}, \ldots)\} \\ &= \tfrac{1}{2} \end{aligned}$$

but $X$ is $T$-inconsistent. As a consequence, we have also the failure of transparency for $T$-equivalence.

$$\hat{p}(\forall x\, A(x)) = 0 \qquad \hat{p}\{A_i | a_i \in D\} = \tfrac{1}{2}$$

even though $\forall x$ A$(x)$ is $T$-equivalent to $\{A_i | a_i \in D\}$.

It should be emphasized here that these difficulties are due entirely to the constraints of individual completeness and the concomitant loss of compactness. It is this loss which renders inapplicable the monotonicity and transparency theorems of section 5 above. If $\Omega$ and $p$ are as defined above, and the logic $T$ is just ordinary predicate logic, without the constraint of individual completeness, then $p$ is a respectable $T$-probability; it is transparent for predicative equivalence and assigns zero to each first order inconsistent subset of $\Omega$. The assignment of a positive value to the set $X$, and of a value to $\forall x\, A(x)$ which differs from that assigned to $\{A(a_i) | a_i \in D\}$, are compatible with this (they entail that the meanings of the quantifiers are not exhausted in their instances) since the first is consistent in first-order logic, and the second two are not first-order equivalent.

If $T$ is an individual complete base logic, then the conjunctive extension of the measure $p$ in a system $\langle \Omega, T, p \rangle$ may, in view of the non-compactness of $T$, fail of $T$-transparency. Such a function would have little interest as a probability. The property of compactness is in a large and general class of cases incompatible with that of individual completeness, and the question thus arises of what other means might be employed to rule out the pathological systems. The response to that question given here is an *ad hoc* one: Constraints are put directly on the measures $p$ and their extensions. This is, for reasons like those discussed with respect to independence, not as worthy a solution from a philosophical point of view as would be a constraint formulated with non-probabilistic condition.

If $\langle \Omega, T, p \rangle$ is a probability system, then the conjunctive extension of $p$ is unique. We say that the system $\langle \Omega, T, p \rangle$ is *standard* if $\hat{p}$ assigns zero to every $T$-inconsistent subset of $\Omega$. If $T$ is compact in $\Omega$ then every system $\langle \Omega, T, p \rangle$ is standard. (Theorem V.2.1–4a.) Even if $T$ is not compact in $\Omega$, so long as $T$ is absolutely consistent and at least tautological there is at least one $p$ such that $\langle \Omega, T, p \rangle$ is standard: Just let $\hat{p}$ assign 1 to every finite subset of some maximal $T$-consistent $\Phi \in M(\Omega)$ and zero to all other finite sets. Then $\hat{p}$ will assign 1 to every subset of $\Phi$ and 0 to all other sets, including $T$-inconsistent sets, and will hence be standard. It is important first to note that the assumption that a system is standard does the work of compactness as far as monotonicity and transparency are concerned. The first crucial theorem here is Theorem V.2.1. In this theorem the fundamental probability characteristics were established for extensions of $T$-measures. Compactness was employed there only in support of two clauses, the first of which

(4a) If $X$ is $T$-inconsistent then $\hat{p}(X) = 0$

is just the assertion that the system is standard, and the second of which

(5b) If $X$ intersects every maximal $T$-consistent subset of $\Omega$ then $\check{p}(X) = 1$

is consequent upon (4a) with no further employment of compactness – (4a) and (1b) are the essential supports. Thus the clauses of theorem V.2.1 hold for standard systems even without the assumption of compactness.

The other major concern is to establish analogues of the monotonicity and transparency theorems of section V.5, without employing compactness. First a

LEMMA A: Let $\langle \Omega, T, p \rangle$ be a standard probability system. Then if the subset $X$ of $\Omega$ $T$-implies the sentence $A$ of $\Omega$,

$$\hat{p}(X) \leqslant \mathrm{p}(A).$$

*Proof:* Let $X = \{A_i, A_2, \ldots\}$ and define the covering nest of $X$ as usual; $X_0 = \wedge$, $X_{i+1} = X_i \cup \{A_{i+1}\}$. assume that $\hat{p}(X) > p(A)$.

For each $X_i \in \{X_i\}$, $p(\wedge X_i) = p(A \wedge \wedge X_i) + p(-A \wedge \wedge X_i)$

and

$$p(A) \geqslant p(A \wedge \wedge X_i) \quad \text{for each } i.$$

Thus,

$$\begin{aligned} \hat{p}(X) &= \inf \{p(\wedge X_i)\} \\ &= \inf \{p(A \wedge \wedge X_i)\} + \inf \{p(-A \wedge \wedge X_i)\} \end{aligned}$$

and

$$p(A) \geqslant \inf \{p(A \wedge \wedge X_i)\}.$$

Thus, $\hat{p}(X)$ being assumed to exceed $p(A)$,

$$\inf\{p(A \wedge \wedge X_i)\} + \inf\{p(-A \wedge \wedge X_i)\} > \inf\{p(A \wedge \wedge X_i)\}$$

so

$$\hat{p}(\{-A\} \cup X) > 0$$

which, since the system is standard, entails that $X \cup \{-A\}$ is $T$-consistent, and, hence, that $X$ does not $T$-imply $A$.

It is now an easy move to the analogues of the monotonicity and transparency results of V.5.

THEOREM 7.1 If $\langle \Omega, T, p \rangle$ is a standard probability system, and the subset $X$ of $\Omega$ $T$-implies the subset $Y$ of $\Omega$, then $\hat{p}(X) \leqslant \hat{p}(Y)$.

*Proof:* If $X$ $T$-implies $Y$ then $X$ $T$-implies every subset of $Y$, hence every finite subset of $Y$. Thus, where $Y_i$ ranges over the finite subsets of $Y$, by the lemma

$$\hat{p}(X) \leqslant p(\wedge Y_i)$$

and hence

$$\hat{p}(X) \leqslant \inf_i p(\wedge Y_i) \leqslant \hat{p}(Y).$$

COROLLARY: If $\langle \Omega, T, p \rangle$ is a standard system, then

(1) $\hat{p}$ is transparent for $T$-equivalence.

(2) If $X$ and $Y$ are subsets of $\Omega$ such that $X$ intersects no maximal $T$-consistent subset of $\Omega$ which is not also intersected by $Y$, then $\check{p}(X) \leqslant \check{p}(Y)$.

*Proofs:* Just as of the corollaries of theorem V.5.5.

The characterization of the probabilities of quantified sentences is now immediate, consequent upon the assumption of a standard and individual complete system.

THEOREM 7.2: Let $\langle \Omega, T, p \rangle$ be a probability system which is standard and predicative, and let $T$ be individual complete in the set $D$ of individual constants occurring in sentences of $\Omega$, then, where $\forall x\, A(x)$ and $\exists x\, A(x)$ are sentences of $\Omega$

(1) $p(\forall x\, A(x)) = \hat{p}\{A(a) | a \in D\}$.

*Proof:* By the second corollary to theorem 7.1 and the $T$-equivalence, in virtue of the individual completeness of $T$, of $\forall x\, A(x)$ and $\{A(a) | a \in D\}$.

(2) $p(\exists x\, A(x) = \check{p}\{A(a) | a \in D\}$.

*Proof:* By the second corollary to theorem 7.1 and the definition of individual completeness.

THEOREM 7.3 (Gaifman condition).[9] Let $\langle \Omega, T, p \rangle$ be a standard and predicative probability system and assume that $T$ is individual complete in the set $D = \{a_1, a_2, \ldots\}$ of individual constants occurring in sentences of $\Omega$.

Let $D_n = \{a_1, \ldots, a_n\}$, for each $n$

$\mathfrak{A}$ = the set of instances of $\forall x\, A(x)$ by individual constants in $D$.

$\mathfrak{A}_n$ = the set of instances of $\forall x\, A(x)$ by individual constants in $D_n$.

Then

(1) $p(\forall x\, A(x)) = \inf\{p(\wedge \mathfrak{A}_n)\}$

(2) $p(\exists x\, A(x)) = \sup\{p(\vee \mathfrak{A}_n)\}$.

*Proof:*

$$\inf\{p(\wedge \mathfrak{A}_n)\} = \inf\{p(\wedge \{A(a) | a \in D_n\})\}$$
$$\sup\{p(\bigvee \mathfrak{A}_n)\} = \sup\{p(\bigvee \{A(a) | a \in D_n\})\}$$

and the theorem follows by theorem 7.2.

The epistemic meaning of this theorem is something like this: A man's belief that everything has a property $S$ is just his conjunctive belief in the infinite set of propositions formed by predicating $S$ of every individual of which (or whom) he can conceive. His belief that something has $S$ is his disjunctive belief in this set. As an analysis of the function of quantifiers in judgment this seems insufficient, for the reason, to be elaborated below, that it derives the quantifiers of universality, and in so doing imbues them with a sort of power to move in and out of judgment. This power is reminiscent of the doctrine which I attributed to Hume, that believing that something is *S* is just believing of something that it is *S*. It is not quite the same, because of the introduction of the concept of disjunctive belief, but it shares the difficulty of that view that it can give no account, for example, of belief that even things of which I presently do not conceive have a character which I do presently comprehend.

The origins of theorem 7.3 and its epistemic consequences should be located not in the concepts of disjunctive and conjunctive belief, but rather in the supposition that the logic of belief conforms to individual completeness. That is to say, in the assumption that it is logically inconsistent to hold that everything that you know of has *S* but that something does not have *S*.

As remarked above, the theorems of this section depend upon a constraint, to standard systems, which is formulated in probabilistic terms, and hence lacks the generality of conditions, such as that of compactness of the base logic, which may be formulated in purely logical terms. The restriction to standard systems is weaker than that of compactness, as the discussion of this section makes clear, but it is not only the strength of conditions that is in question. What are wanted are insights into the connections of logical concepts and those of probability, or partial belief, and the more one is forced to restrict the concepts of belief and probability in their own terms the less general are those insights.

The restriction to standard systems raises also the question of its justification. It is clear that a man's beliefs might conform to the laws of predicative and individual complete probability as far as sentences and finite sets of them are concerned, and yet fail to be standard. He might still believe some infinite inconsistent (in terms of first-order, individual complete, logic) set of sentences to positive degree. That is just the problem raised by logics which are not compact; probability functions on finite sets may have extensions which are non-probabilistic.

Isolation of the class of standard systems has no apparent justification in terms of coherence, for we have no account of bets on infinite sets of propositions, or of infinite sets of bets, and many of the same problems arise as were mentioned in the discussion of denumerable additivity.

We do, however, have at least a partial answer to this question in terms of transparency. Obviously, if $p$ is $T$-transparent, then $\langle \Omega, T, p \rangle$ is standard. So, recalling the previous theorem and its corollaries, the restriction to standard systems is in a sense equivalent to that to finite probabilities which have $T$-transparent extensions. That gives the restrictions to standard systems a base in requirements of reasonableness.

## NOTES

[1] The reader unfamiliar with algebra may consult the appendix on set theory and Boolean algebra for a brief discussion.

[2] See theorem III.1.4.

[3] *Hume*, p. 125.

[4] See, for example, section 10 of *Carnap, Logical Foundations*.

[5] See, for example, *Savage*, Chapter 3, section 4. Carnap, in essay 2 of *Carnap* 1971 (definition 1.6) also picks the broader definition. See Jeffrey's remarks on p. 220 of the same volume.

[6] As does the equivalent notion of exchangeable events. See *De Finetti*.

[7] Section 35.

[8] In *Russell*, 1956.

[9] Cf. *Gaifman*, and *Scott and Krauss*. The present development differs from these in using sets of sentences rather than infinite formulas.

CHAPTER VI

# INFINITY AND THE SUM CONDITION

## VI.1. Generalization of the Sum Condition

In Chapter IV the laws of probability were related to the simple sum condition:

If $\Omega$ is a denumerable field of sentences, $T$ an absolutely consistent and at least tautological logic, and $p$ a numerical function on $\Omega$, then $\Omega$, $T$ and $p$ satisfy the simple sum condition if and only if for every subset $X$ of $\Omega$.

(i) If the sum of $p$ over $X$ is finite, then there are maximal $T$-consistent $\Phi_0$ and $\Phi_1 \in M(\Omega)$ such that

$$N(\Phi_0 \cap X) \leqslant \sum_{A \in X} p(A) \leqslant N(\Phi_1 \cap X).$$

(ii) If the sum of $p$ over $X$ is infinite, then there is no finite upper bound on the quantities $N_j(X)$. I.e., for every finite $n$ there is some $\Phi_j \in M(\Omega)$ such that $n < N_j(X)$.

Combining the theorems IV.2.1 and IV.3.3 we have

THEOREM 1.1. Let $\Omega$ be a denumerable field of sentences, $T$ be an absolutely consistent and at least tautological logic, and assume that $T$-inconsistency is compact in $\Omega$. Then $\langle \Omega, T, p \rangle$ is a probability system if and only if $\langle \Omega, T, p \rangle$ satisfies the simple sum condition.

In the preceding chapter, V, the functions $\hat{p}$ and $\check{p}$, extensions of a probability $p$ on sentences, were defined for sets of sentences and were shown to be probabilistic in a straightforward sense. That was the import of Theorem V.2.1. In the present chapter the sum condition is formulated to apply to collections of sets of sentences, and its equivalence to the laws of probability is established for this extended sense as well.

The first step is an exercise in definition. The plausibility of the simple sum condition as formulated in chapter IV rests on its relation to the concept of coherence and to the laws of probability. The question now is how that condition should be interpreted or extended so as to have application to functions defined on sets of sentences. We look first at this question for conjunctive extensions.

Consider, first, the conjunctive extension $\hat{p}$ of a probability $p$ as it applies to finite subsets of a field $\Omega$. For such sets the values of $\hat{p}$ are just those assigned by $p$ to the conjunctions of the finite sets

$$\hat{p}(X) = p(\wedge X).$$

Now let $\chi = \{X_1, \ldots\}$ be a collection of finite subsets of $\Omega$. If $p$ satisfies the sum condition on $\Omega$, then there are maximal $T$-consistent $\Phi_0$ and $\Phi_1$, members of $M(\Omega)$, such that

$$\text{(i)} \qquad N[\Phi_0 \cap \{\wedge X_1, \ldots\}] \leqslant \sum_{i=1}^{\infty} p(\wedge X_i) \leqslant N[\Phi_1 \cap \{\wedge X_i, \ldots\}].$$

This expression may be simplified by remarking that for any maximal $T$-consistent $\Phi$, $N[\Phi \cap \{\wedge X_1, \ldots\}]$ is just the number of the sets $X_i$ which are subsets of $\Phi$ (allowing ourselves for the moment to think of a denumerable infinity as a number). Then (i) may be reformulated;

$$\text{(ii)} \qquad N\{X_i | X_i \in \chi \wedge X_i \subseteq \Phi_0\} \leqslant \sum_{i=1}^{\infty} p(\wedge X_i) \leqslant N\{X_i | X_i \in \chi \wedge X_i \subseteq \Phi_1\}.$$

It will be convenient to introduce the notation

$$\hat{N}_j(\chi)$$

to abbreviate

$$N\{X_i | X_i \in \chi \wedge X_i \subseteq \Phi_j\}.$$

That is, where $j$ indexes the maximal $T$-consistent subsets of $\Omega$, $\hat{N}_j(\chi)$ is the number of the members of $\chi$ which are subsets of the maximal $T$-consistent $\Phi_j$. Then, $\chi$ remaining a collection of finite subsets of $\Omega$, the definition of $p$ yields as a further formulation, yet simpler,

$$\text{(iii)} \qquad \hat{N}_0(\chi) \leqslant \sum_{X \in \chi} \hat{p}(X) \leqslant \hat{N}_1(\chi)$$

and, recalling the notation of chapter V, the sum condition as applied to collections of finite subsets of $\Omega$ may be put

(iv) $$\min_j \hat{N}_j(\chi) < \sum_{X \in \chi} \hat{p}(X) \leqslant \max_j \hat{N}_j(\chi).$$

So far no discretion in definition has been called for: (iv) is a consequence of the sum condition as formulated in chapter V so long as $\chi$ consists of finite subsets of $\Omega$, in the case in which the sum of $\hat{p}$ over these sets is finite. The epistemic meaning of (iv) is something like this. The sum of the conjunctive probabilities of a collection of finite sets must lie between (a) the lowest number of them which may be consistently conjunctively asserted, without conjunctively asserting any other, and (b) the highest number of them which may be consistently asserted. The extended sum condition retains this sense and applies it to the conjunctive assertion of infinite sets of sentences:

If $\Omega$ is a denumerable field of sentences, $T$ an absolutely consistent and at least tautological logic, and $\hat{p}$ any numerical function defined on subsets of $\Omega$, then $\hat{p}$ is said to conform to the *conjunctive sum condition* if for every denumerable collection $\chi$ of subsets of $\Omega$,

(i) If the sum of $p$ over $\chi$ is finite, then there are $\Phi_0$ and $\Phi_1 \in M(\Omega)$ such that

$$\hat{N}_0(\chi) \leqslant \sum_{X \in \chi} \hat{p}(X) \leqslant \hat{N}_1(\chi).$$

(ii) If the sum of $p$ over $\chi$ is infinite, then there is no finite upper bound on the quantities $\hat{N}_j(\chi)$. I.e; for every finite $n$ there is some $\Phi_j$, such that

$$\hat{N}_j(\chi) > n.$$

The definitional problem for the disjunctive extension is quite similar. Let $\chi = \{X_1, \ldots\}$ be a collection of finite subsets of $\Omega$. Then if $p$ satisfies the simple sum condition as formulated in chapter V, for members of $\Omega$, there are maximal $T$-consistent subsets of $\Omega$, $\Phi_0$ and $\Phi_1$ such that

(i) $$N[\Phi_0 \cap \{\vee X_1, \ldots\}] \leqslant \sum_{i=1}^{\infty} p(\vee X_i) \leqslant N[\Phi_1 \cap \{\vee X_1, \ldots\}].$$

For any maximal $T$-consistent $\Phi$

$N[\Phi \cap \{\vee X_1, \ldots\}]$ is just the number of the disjunctions $\vee X_i$ which are members of $\Phi$, and a disjunction is a member of a maximal $T$-consistent set just in case at least one of its disjuncts is a member of the set. That is to say, $\vee X_i$ is a member of $\Phi$ just in case $\Phi \cap X_i$ is not null. Thus, again assuming the members of $\chi$ to be finite, (i) may be reformulated,

$$\text{(ii)} \qquad N\{X_i | X_i \in \chi \wedge (X_i \cap \Phi_0) \neq \Lambda\} \leqslant \sum_{i=1}^{\infty} p(\vee X_i)$$
$$\leqslant N\{X_i | X_i \in \chi \wedge (X_i \cap \Phi_1) \neq \Lambda\}.$$

If we now introduce the notation

$$\check{N}_j(\chi)$$

to abbreviate

$$N\{X_i | X_i \in \chi \wedge (X_i \cap \Phi_j) \neq \Lambda\}$$

and make use of the definition of the disjunctive extension for finite sets, we have,

$$\text{(iii)} \qquad \check{N}_0(\chi) \leqslant \sum_{X \in \chi} \check{p}(X) \leqslant \check{N}_1(\chi).$$

Finally, the sum condition as applied to collections of finite subsets of $\Omega$ yields,

$$\text{(iv)} \qquad \min_j \check{N}_j(\chi) \leqslant \sum_{X \in X} \check{p}(X) \leqslant \max_j \check{N}_j(\chi).$$

The epistemic meaning of which is roughly that the sum of the disjunctive probabilities of a number of finite sets must lie between (a) the least number of them which may be consistently disjunctively asserted without asserting any others, and (b) the highest number of them which may be consistently asserted. As in the conjunctive case, the import of this is retained in the extension to functions defined for infinite sets:

If $\Omega$, $T$, are as in the previous definition and $\check{p}$ is any numerical function defined on subsets of $\Omega$, then $\check{p}$ is said to conform to the *disjunctive sum condition* if for every denumerable collection $\chi$ of subsets of $\Omega$

(i) If the sum of $\check{p}$ over $\chi$ is finite then there are $\Phi_0$ and $\Phi_1 \in M(\Omega)$ such that

$$\check{N}_0(\chi) \leqslant \sum_{X \in \chi} \check{p}(X) \leqslant \check{N}_i(\chi).$$

(ii) If the sum of $p$ over $\chi$ is infinite, then there is no finite upper bound on the quantities $\check{N}_j(\chi)$, i.e., for every finite $n$, there is some $\Phi_j$ such that $n < \check{N}_j(\chi)$.

Clearly, if $\hat{p}$ (or $\check{p}$) is a numerical function defined on subsets of a field $\Omega$ which satisfies the conjunctive (respectively disjunctive) sum condition, then the function $p$ defined for members of the field

$$p(A) = \hat{p}\{A\}$$

or, equivalently

$$p(A) = \check{p}\{A\}$$

satisfies the simple sum condition. And, as the preceding discussion shows, if $p$ satisfies the simple sum condition for members of $\Omega$, then the functions

$$\hat{p}(X) = p(\bigwedge X)$$
$$\check{p}(X) = p(\bigvee X)$$

satisfy the conjunctive and disjunctive sum conditions respectively for finite subsets of $\Omega$. Thus these definitions conform to the earlier one, and are extensions of it.

Before investigating the relations of the generalized sum conditions to the laws of probability for $\check{p}$ and $\hat{p}$, it will be useful to develop some consequences about the cardinalities $\hat{N}_j(\chi)$ and $\check{N}_j(\chi)$. In view of the duality of conjunction and disjunction and the classical nature of negation in tautological logics $T$, these numbers (or cardinalities) will be related in ways which then – through the sum conditions – relate conjunctive and disjunctive probabilities. Perhaps the most important and obvious relation of conjunctive and disjunctive probabilities; when the base logic is at least tautological, is afforded by De Morgan's laws, that is, by the equivalence of

$$-(A \vee B) \quad \text{to} \quad -A \wedge -B$$

and of

$$-(A \wedge B) \quad \text{to} \quad -A \vee -B.$$

The consequences for tautological probabilities is that

$$p(A \vee B) = 1 - p(-A \wedge -B)$$
$$p(A \wedge B) = 1 - p(-A \vee -B)$$

etc.

The generalization of this to finite conjunctions and disjunctions of greater length is not hard to come by. In the present context, however, we are interested in basing these relations upon the sum condition. The present theorem is a simple and intuitive step in this direction. It should be remarked that it applies to *finite* collections of *infinite* sets of sets of sentences. A little care may be required to keep the import of $\bar{X}$ distinct from that of $\bar{\chi}$.

THEOREM 1.2. Let $\Omega$ be a denumerable field of sentences, and $T$ an absolutely consistent and at least tautological logic on $\Omega$. For each $X \subseteq \Omega$, let $\bar{X}$ be the set of negations of sentences in $X$:

$$\bar{X} = \{-A \mid A \in X\}$$

and for each collection $\chi$ of subsets of $\Omega$, let

$$\bar{\chi} = \{\bar{X} \mid X \in \chi\}$$

then, if $\chi$ is a finite collection of (perhaps infinite) subsets of $\Omega$, for each maximal $T$-consistent $\Phi_j \in M(\Omega)$

$$(1) \qquad \hat{N}_j(\bar{\chi}) = N(\chi) - \check{N}_j(\chi)$$
$$(2) \qquad \check{N}_j(\bar{\chi}) = N(\chi) - \hat{N}_j(\chi).$$

*Proof*: Consider a set $X \in \chi$ and assume that $\bar{X} \subseteq \Phi_j$. Then for each $A \in X$, $-A \in \Phi_j$. So if $A \in X$, then $A \notin \Phi_j$. Thus $X \cap \Phi_j = \Lambda$ and thus

$$\{X \mid X \in \chi \wedge \bar{X} \subseteq \Phi_j\} \subseteq \{X \mid X \in \chi \wedge X \cap \Phi_j = \Lambda\}.$$

Similarly, if $X \cap \Phi_j = \Lambda$, then for every $A \in X$, $-A \in \Phi_j$ and $X \subseteq \Phi_j$. Thus

$$\text{(i)} \qquad \{X \mid X \in \chi \wedge \bar{X} \subseteq \Phi_j\} = \{X \mid X \in \chi \wedge X \cap \Phi_j = \Lambda\}.$$

A symmetrical argument establishes

$$\text{(ii)} \qquad \{X \mid X \in \chi \wedge \bar{X} \cap \Phi_j \neq \Lambda\} = \{X \mid X \in \chi \wedge X \nsubseteq \Phi_j\}$$

and since

$$\chi = \{X \mid X \in \chi \wedge \bar{X} \subseteq \Phi_j\} \cup \{X \mid X \in \chi \wedge \bar{X} \nsubseteq \Phi_j\}$$

we have

$$\{X \mid X \in \chi \wedge X \subseteq \Phi_j\} = \chi - \{X \mid X \in \chi \wedge X \cap \Phi_j \neq \Lambda\}$$

since $\chi$ is finite

$$N[\chi - \{X \mid X \in \chi \wedge X \cap \Phi_j \neq \Lambda\}] =$$
$$= N(\chi) - N\{X \mid X \in \chi \wedge X \cap \Phi_j \neq \Lambda\}.$$

So

$$N\{X \mid X \in \chi \wedge \overline{X} \subseteq \Phi_j\} = N(\chi) - N\{X \mid X \in \chi \wedge X \cap \Phi_j \neq \Lambda\}$$

and

$$\hat{N}_j(\overline{\chi}) = N(\chi) - \check{N}_j(\chi).$$

Thus, (1).

(2) follows similarly from (ii).

We have also an obvious and useful

THEOREM 1.3. Let $\Omega$ and $T$ be as in the previous theorem, and let $\overline{X}$ and $\overline{\chi}$ be defined as there. Then if $\chi$ is denumerably infinite and either of $\hat{N}_j(\overline{\chi})$, $\check{N}_j(\chi)$ is finite, the other is denumerably infinite. If $\chi$ is denumerably infinite and either of $\check{N}_j(\overline{\chi})$, $\hat{N}_j(\chi)$ is finite, then the other is denumerably infinite.

## VI.2. Probability Entails the Generalized Sum Conditions

In this section the theorems of chapter V which show that all probability systems satisfy the simple sum condition are extended to show that the generalized conditions hold for conjunctive and disjunctive extensions. The major theorem of this section is

THEOREM 2.1 Let $\langle \Omega, T, p \rangle$ be a compact probability system, and let $\chi = \{X^1, \ldots\}$ be a denumerable collection of (possibly infinite) subsets of $\Omega$. Then, where $\hat{p}$ is the conjunctive extension of $p$, and $\hat{N}_j$ is as defined above:

(1) If $\sum_{m=1}^{k} \hat{p}(X^m)$ approaches a limit as $k$ increases without bound, then there are maximal $T$-consistent $\Phi_0$ and $\Phi_1$, members of $M(\Omega)$, such that

$$\hat{N}_0(\chi) \leqslant \lim_{k \to \infty} \sum_{m=1}^{k} \hat{p}(X^m) \leqslant \hat{N}_1(\chi)$$

(2) If $\sum_{m=1}^{k} \hat{p}(X^m)$ increases without bound as $k$ increases without bound, then for every finite quantity $n$, there is some $\Phi_j \in M(\Omega)$, such that $n < \hat{N}_j(\chi)$.

The argument for this theorem will refer throughout to an *array* of subsets of $\Omega$ which we now describe. Let $\chi = \{X^1, \ldots\}$ be as described in the theorem, and for each member $X^m$ of $\chi$ let $\{X_i^m\}$ be a covering nest of $X^m$. Thus, for each $m$

$$\Lambda = X_0^m \subseteq X_1^m \subseteq \cdots \subseteq X_1^m \subseteq \cdots \subseteq X^m$$

and for each $i$, let $\chi_i$ be the collection of the $i$'th members of the respective covering nests;

$$\chi_i = \{X_i^1, X_i^2, \ldots, X_i^m, \ldots\}.$$

The array of subsets to which we refer consists of the collections

$$\begin{array}{cccccc}
\chi_1 & \chi_2 & \cdots & \chi_i & \cdots & \chi \\
X_1^1 & X_2^1 & \ldots & X_1^1 & \ldots & X^1 \\
\vdots & \vdots & & \vdots & & \vdots \\
X_1^j & X_2^j & \ldots & X_i^j & \ldots & X^j \\
\vdots & \vdots & & \vdots & & \vdots \\
X_1^m & X_2^m & \ldots & X_i^m & \ldots & X^m \\
\vdots & \vdots & & \vdots & & \vdots
\end{array}$$

It may be helpful to consider briefly some features of this array and of a function $\hat{p}$ which we may assume defined for its members: Each row of the array is a covering nest of its right-most member. Thus $\hat{p}(X_i^m)$ is monotonically non-increasing as $i$ increases, for a fixed row $m$. Thus, for each $k$, $\sum_{m=1}^k \hat{p}(X_i^m)$ is monotonically non-increasing as $i$ increases. The left-most column of the array consists of the zero'th members of the respective nests, each of these is just the null set. Thus as $k \to \infty$, $\sum_{m=1}^k \hat{p}(X_0^m)$ increases without bound, for at each $k$,

$$\sum_{m=1}^{k} \hat{p}(X_0^m) = k.$$

If, however, for some column $i_0$

$$\sum_{m=1}^{k} \hat{p}(X_{i_0}^m)$$

approaches a finite limit as $k$ increases without bound, then this limit may be identified with the sum of $\hat{p}$ for members of the column $i_0$

$$\sum_{m=1}^{\infty} \hat{p}(X_{i_0}^m) = \lim_{k \to \infty} \sum_{m=1}^{k} \hat{p}(X_{i_0}^m)$$

and we shall have, in this case, that, for all $i \geqslant i_0$,

$$\sum_{m=1}^{\infty} \hat{p}(X^m) \leqslant \sum_{m=1}^{\infty} \hat{p}(X_i^m) \leqslant \sum_{m=1}^{\infty} \hat{p}(X_{i_0}^m).$$

In this way, we shall be able to approach $\sum_{m=1}^{\infty} \hat{p}(X^m)$ from above, and it is just this feature that permits exploitation of the finite form of the theorem to find the maximal $T$-consistent $\Phi_1$ of clause (1) of the theorem.

$\sum_{m=1}^{\infty} \hat{p}(X^m)$, assuming it to exist, may also be approached from *below* by the sums, for successive $k$

$$\sum_{m=1}^{k} \hat{p}(X^m)$$

and it is this which lets us find the $\Phi_0$ of clause (1).

We commence the discussion of the theorem with a lemma in which some of these matters are treated more clearly, and in which some limits and infima are related.

LEMMA A. If $\sum_{m=1}^{k} \hat{p}(X^m)$ approaches a finite limit as $k$ increases without bound, then there is some $i_0$ such that for all $i \geqslant i_0$, $\sum_{m=1}^{k} \hat{p}(X_i^m)$ approaches a finite limit as $k$ increases without bound.

Thus

$$\text{(i)} \qquad \inf_i \left\{ \lim_{k \to \infty} \sum_{m=1}^{k} \hat{p}(X_i^m) \right\}$$

exists, and further,

$$\text{(ii)} \qquad \sum_{m=1}^{\infty} \hat{p}(X^m) = \lim_{k \to \infty} \sum_{m=1}^{k} \hat{p}(X^m) = \lim_{k \to \infty} \sum_{m=1}^{k} \inf_i \left\{ \hat{p}(X_i^m) \right\}$$

$$= \inf_i \left\{ \lim_{k \to \infty} \sum_{m=1}^{k} \hat{p}(X_i^m) \right\}$$

*Proof* (*i*): Assume that for every $i$, $\sum_{m=1}^{k} \hat{p}(X_i^m)$ increases without bound as $k$ increases without bound. Then

$$\inf_i \left\{ \sum_{m=1}^{k} \hat{p}(X_i^m) \right\}$$

(which is, of course, always defined) also increases without bound as $k$ increases without bound. Hence

$$\sum_{m=1}^{k} \inf_i \left\{\hat{p}(X_i^m)\right\} = \sum_{m=1}^{k} \hat{p}(X^m)$$

increases without bound as $k \to \infty$.

*Proof* (*ii*): Since

$$\lim_{k\to\infty} \sum_{m=1}^{k} \hat{p}(X^m)$$

exists, this limit is just the infinite sum

$$\sum_{m=1}^{\infty} \hat{p}(X^m).$$

$\hat{p}(X^m)$ is by definition $\inf\{\hat{p}(X_i^m)\}$, so

$$\sum_{m=1}^{k} \hat{p}(X^m) = \sum_{m=1}^{k} \inf_i \{\hat{p}(X_i^m)\}$$

and these expressions thus maintain their identity in the limit, which is assumed to exist in the hypothesis of the lemma.

*Proof* (*iii*): Notice first that for each $k$

$$\sum_{m=1}^{k} \inf_i \{\hat{p}(X_i^m)\} = \inf_i \left\{ \sum_{m=1}^{k} \hat{p}(X_i^m) \right\}$$

we have already seen that the left side approaches a finite limit as $k \to \infty$, thus

$$\text{(iv)} \qquad \lim_{k\to\infty} \sum_{m=1}^{k} \inf_i \{\hat{p}(X_i^m)\} = \lim_{k\to\infty} \inf_i \left\{ \sum_{m=1}^{k} \hat{p}(X_i^m) \right\}.$$

Clearly for each $i$

$$\lim_{k\to\infty} \inf_i \left\{ \sum_{m=1}^{k} \hat{p}(X_i^m) \right\} \leqslant \lim_{k\to\infty} \sum_{m=1}^{k} \hat{p}(X_i^m)$$

so the expression

$$\alpha\colon \lim_{k\to\infty} \inf_i \left\{ \sum_{m-1}^{k} \hat{p}(X_i^m) \right\}$$

is a lower bound on the set

$$\beta: \left\{ \lim_{k \to \infty} \sum_{m=1}^{k} \hat{p}(X_i^m) \right\}.$$

By (i) above the infimum of $\beta$ exists, and for each $k$, for sufficiently large $i$,

$$\sum_{m-1}^{k} \hat{p}(X_i^m)$$

may be made to approach

$$\inf_i \left\{ \sum_{m-1}^{k} \hat{p}(X_i^m) \right\}$$

as closely as desired. Thus if a quantity $n$ exceeds $\alpha$, $n$ must also exceed some member of $\beta$. Thus $\alpha$ is the greatest lower bound on the set $\beta$. And hence

$$\lim_{k \to \infty} \inf_i \left\{ \sum_{m=1}^{k} \hat{p}(X_i^m) \right\} = \inf_i \left\{ \lim_{k \to \infty} \sum_{m=1}^{k} \hat{p}(X_i^m) \right\}$$

which with (iv) entails (iii).

We now state and prove a lemma which describes an important logical feature of the sub-arrays consisting of finitely many rows of the full array:

LEMMA B. For each $k$ let

$$\chi^k = \{X^1, \ldots, X^k\}$$

and for each $i$ and $k$ let

$$\chi_i^k = \{X_1^1, \ldots, X_i^k\}$$

then for each $\chi^k$ there is some $\chi_i^k$ such that

$$\max_j \hat{N}_j(\chi_k) = \max_j \hat{N}_j(\chi_i^k).$$

*Proof:* Since every $X_i^m$ is a subset of the corresponding $X^m$, for every $\Phi_j$, $\hat{N}(\chi^k) \leqslant \hat{N}_j(\chi_i^k)$ for every $i$ and $k$.

Thus

$$\max \hat{N}_j(\chi^k) \leqslant \max_j \hat{N}_j(\chi_i^k)$$

for every $i$ and $k$, and we need only show that, given a set $\chi^k$, there is some $i^*$ such that

$$\max N_j(\chi_{i^*}^k) \leqslant \max_j N_j(\chi^k).$$

Let $\max_j N_j(\chi^k)=r$. Then no maximal $T$-consistent $\Phi$ includes more than $r$ of the sets $\{X^1,\ldots, X^k\}$. Let $\alpha_1,\ldots, \alpha_s$ (where $s=(k/(r+1)))$ be all the $r+1$ membered subsets of $\chi^k=\{X^1,\ldots, X^k\}$. For each $\alpha_n$ and $\chi_i^k$, let $\alpha_n(i)$ be the corresponding $r+1$ membered subset of $\chi_i^k$. (If, for example, $r=2$ and $\alpha_n=\{X^2, X^3, X^8\}$, then for each $i$, $\alpha_n(i)=\{X_i^2, X_i^3, X_i^8\}$). Thus, each $\alpha_n$ may be thought of as selecting $r+1$ rows from the array, and the argument $i$, or absence of it, as selecting a column of the array. Since $\alpha_1,\ldots, \alpha_s$ are all the $r+1$ membered subsets of $\chi^k$, for each $i$, $\alpha_1(i),\ldots, \alpha_s(i)$ are all the $r+1$ membered subsets of $\chi_i^k$.

Since no maximal $T$-consistent $\Phi$ includes more than $r$ of $X^1,\ldots, X^k$, each union $\bigcup\alpha_n$, for $1\leqslant n\leqslant s$, is $T$-inconsistent, and since $T$-inconsistency is compact in $\Omega$, each union $\bigcup\alpha_n$ has a finite $T$-inconsistent subset. Thus, in view of the structure of the covering nests $\{X_i^m\}$, for each $n$ there is some (column of the array) $i(n)$ such that $\bigcup\alpha_n^{i(n)}$ is $T$-inconsistent. Again in view of the structure of the covering nests, we have that

(i) For each $\alpha_n$ there is some $i(n)$ such that for all $i\geqslant i(n)$, $\bigcup\alpha_n^i$ is $T$-inconsistent.

Now let $i^*$ be the maximum of the values $i(1),\ldots, i(s)$. Then for each $n$, $1\leqslant n\leqslant s$, $\bigcup\alpha_n^{i^*}$ is $T$-inconsistent. Thus every $r+1$ membered subset of $\chi_{i^*}^k$ is $T$-inconsistent, and no maximal $T$-consistent $\Phi$ includes more than $r$ of $X_{i^*}^1,\ldots, X_{i^*}^k$. Thus

$$\max_j \hat{N}_j(\chi_{i^*}^k) \leqslant r = \max_j \hat{N}_j(\chi^k)$$

which completes the proof of lemma B.

One more brief lemma will permit a smooth proof of the main theorem:

LEMMA C. Where $\chi^k$ is defined as in lemma (B), for each $k$; For each $k$ there is some maximal $T$-consistent $\Phi_j$ such that

$$\sum_{m=1}^{k} \hat{p}(X^m) \leqslant \hat{N}_j(\chi^k).$$

*Proof*:

$$\sum_{m=1}^{k} \hat{p}(X^m) = \sum_{m=1}^{k} \inf_i \{\hat{p}(X_i^m)\} = \inf_i \left\{ \sum_{m=1}^{k} \hat{p}(X_i^m) \right\}.$$

Now assume that for all $\Phi_j$

$$\hat{N}_j(\chi^k) < \inf_i \left\{ \sum_{m=1}^{k} \hat{p}(X^m) \right\}$$

then for all $k$

$$\hat{N}_j(\chi^k) < \inf_i \left\{ \sum_{m=1}^{k} \hat{p}(X^m) \right\}$$

and

$$\max_j \hat{N}_j(\chi_{i^*}^k) < \sum_{m=1}^{k} \hat{p}(X_i^m).$$

for all $i$. Thus by lemma (B), for some $i^*$,

$$\max_j \hat{N}_j(\chi_{i^*}^k) < \sum_{m=1}^{k} \hat{p}(X_{i^*}^m).$$

But this contradicts theorem VI.1.1, since the sets $X_i^m$ are finite.

COROLLARY. For every $k$ there is some $\Phi_j$ such that $\check{N}_j(\chi) \leqslant \sum_{m=1}^{k} \check{p}(X^m)$.

Now the proof of theorem 1.3 is quite simple:

Clause (1): Assume that $\sum_{m=1}^{k} \hat{p}(X^m)$ approaches a limit as $k$ increases without bound. We argue first for the existence of a maximal $T$-consistent $\Phi_1$ such that

$$\sum_{m=1}^{\infty} \hat{p}(X^m) = \lim_{k\to\infty} \sum_{m=1}^{k} \hat{p}(X^m) \leqslant \hat{N}_1(\chi).$$

To prove this assume the contrary, that is, that

(i) For every $\Phi_j$,

$$\hat{N}_j(\chi) < \lim_{k\to\infty} \sum_{m=1}^{k} \hat{p}(X^m)$$

then, for some $k$, for every $\Phi_j$

$$\hat{N}_j(\chi) < \sum_{m=1}^{k} \hat{p}(X^m).$$

For each $k$, $\hat{N}_j(\chi^k) \leqslant \hat{N}_j(\chi)$, so for some $k$, for every $\Phi_j$, $\hat{N}_j(\chi^k) < \sum_{m=1}^{k} \hat{p}(X^m)$ in contradiction of lemma (C). Thus (i) must be rejected and we have that for some maximal $T$-consistent $\Phi_1$,

$$\sum_{m-1}^{\infty} \hat{p}(X^m) \leqslant \hat{N}_1(\chi).$$

To complete the proof of clause (1) it remains to show that under the assumption of that clause there is some maximal $T$-consistent $\Phi_0$, such that

$$\hat{N}_0(\chi) \leqslant \sum_{m=1}^{\infty} \hat{p}(X^m).$$

To prove this, we remark first that by lemma A for some $\chi_i$, $\sum_{m=1}^{\infty} \hat{p}(X_i^m)$ is finite. Thus by theorem 1.1, on denumerable collections of finite sets, for each $\chi_i$, there is some $\Phi_j$ such that

$$\text{(i)} \qquad \hat{N}_j(\chi_i) \leqslant \sum_{m=1}^{\infty} \hat{p}(X_i^m)$$

i.e., such that

$$\hat{N}_j(\chi_i) \leqslant \lim_{k\to\infty} \sum_{m=1}^{k} \hat{p}(X_i^m).$$

By lemma (Aii).

$$\sum_{m=1}^{\infty} \hat{p}(X^m) = \inf_i \left\{ \lim_{k\to\infty} \sum_{m=1}^{k} \hat{p}(X_i^m) \right\}.$$

Now if

(ii) For every $\Phi_j$, $\sum_{m=1}^{\infty} \hat{p}(X^m) < \hat{N}_j(\chi)$, then for every $\Phi_j$

$$\inf_i \left\{ \lim_{k\to\infty} \sum_{m=1}^{k} \hat{p}(X_i^m) \right\} < \hat{N}_j(\chi)$$

and for some $i_0$, for every $\Phi_j$

$$\lim_{k\to\infty} \sum_{m=1}^{k} \hat{p}(X_{i_0}^m) < \hat{N}_j(\chi).$$

Since $\hat{N}_j(\chi) \leqslant \hat{N}_j(\chi_i)$, for each $i$, we have that for some $i_0$, for every $\Phi_j$.

$$\sum_{m=1}^{\infty} \hat{p}(X_{i_0}^m) < \hat{N}_j(\chi_i).$$

In contradiction of (i). Thus we must deny (ii), and clause (1) is established.

COROLLARY: For every $k$, there is some $\Phi_j$ such that

$$\sum_{m=1}^{k} \hat{p}(X^m) \leqslant \hat{N}_j(\chi^k).$$

Clause (2). Assume that $\sum_{m=1}^{k} \hat{p}(X^m)$ increases without bound as $k \to \infty$. Then for every finite quantity $n$, there is some $k$ such that

$$n < \sum_{m=1}^{k} \hat{p}(X^m).$$

So – by lemma (C) – for every $n$ there is some $k$ such that

$$n < \max_j \hat{N}_j(\chi_k)$$

and, since for every maximal $T$-consistent $\Phi_j$ and every $k$

$$\hat{N}_j(\chi^k) \leqslant \hat{N}_j(\chi)$$

for every $n$ there is some $\Phi_j$ such that

$$n < \hat{N}_j(\chi).$$

Which completes the proof of clause (2) and the theorem.

The next theorem is the analogue of theorem 2.1 for disjunctive extensions. Its proof is, though not quite obvious, greatly facilitated by the disjunctive corollaries established so far. One additional logical lemma, referring to the right column of the array used in the above arguments, is also a help.

LEMMA D. Let $\langle \Omega, T, p \rangle$ and $\chi$ be as in the preceding theorem, let $\check{N}_j$ be as defined above, and, where $\chi = \{X^1, \ldots\}$, for each $k$ let $\chi^k = \{X^1, \ldots, X^k\}$. Then if for every $k$ there is some maximal $T$-consistent $\Phi_j$ such that $\check{N}_j(\chi^k) \leqslant r$, then for some $\Phi_j$, $\check{N}_j(\chi) \leqslant r$.

The assumption that $T$-inconsistency is compact will figure importantly in this argument. For each $n$ let an *n-membered selection* from $\chi$, or from $\chi^k$ if $k \geqslant n$, be a conjunction of some $n$ sentences, each belonging to a different $X^m \in \chi$ (respectively $\chi^k$). So every conjunct of an $n$-membered selection is from some member of $\chi$ (respectively $\chi^k$) and no two conjuncts are from the same set.

Now, on the assumption of the lemma, for each $\chi^k = \{X^1, \ldots, X^k\}$, there is some maximal $T$-consistent $\Phi_j$ which intersects no more than $r$ of $\{X^1, \ldots, X^k\}$. Thus if $A = (A_1 \wedge \cdots \wedge A_{r+1})$ is an $r+1$ membered selection from $\chi^k$, $A$ is not a member of this $\Phi_j$, and, since $\Phi_j$ is maximal, $-A$, the negation of $A$, is a member of $\Phi_j$. Thus for every $\chi^k$ there is some maximal $T$-consistent $\Phi_j$ which includes the negation of each $r+1$ membered selection from $\chi^k$. Thus for each $\chi^k$, the collection of all negations of $r+1$ membered selections from $\chi^k$ is $T$-consistent.

Let $\Gamma$ be the set of all negations of $r+1$ membered selections from $\chi$, and let $\Gamma'$ be any finite subset of $\Gamma$. Then there is some $\chi^k$ such that every member of $\Gamma'$ is the negation of an $r+1$ membered selection from $\chi^k$, hence $\Gamma'$ is $T$-consistent. Thus every finite subset of $\Gamma$ is $T$-consistent, and, since $T$-inconsistency is compact, $\Gamma$ is $T$-consistent. So for some maximal $T$-consistent $\Phi_j$, $\Gamma \subseteq \Phi_j$. If $A$ is any $r+1$ membered selection from $\chi$, $-A \in \Phi_j$, so $A \notin \Phi_j$. Thus $\Phi_j$ can intersect at most $r$ members of $\chi$ and $\check{N}_j(\chi) \leqslant r$. This establishes the lemma.

The argument for theorem 2.2 will also depend upon the previously established corollaries of the preceding theorem and its lemmas. The presuppositions of that theorem and the description of the array are assumed. We mention the corollaries here for reference.

COROLLARY TO LEMMA C. For every $\chi^k$ there is some maximal $T$-consistent $\Phi_j$ such that

$$\check{N}_j(\chi^k) \leqslant \sum_{m=1}^{k} \check{p}(X^m).$$

COROLLARY TO THE THEOREM. For every $\chi^k$ there is some maximal $T$-consistent $\Phi_j$ such that

$$\sum_{m=1}^{k} \check{p}(X^m) \leqslant \check{N}_j(\chi^k).$$

We may also remark, for reference below, that

($E$) If $\max_j \check{N}_j(\chi)$ is finite, then for every $\chi^k$

$$\max_j \check{N}_j(\chi^k) \leqslant \max_j \check{N}_j(\chi).$$

Now to the

THEOREM 2.2 Let $\langle \Omega, T, p \rangle$ be a compact probability system and let $\chi = \{X^1, \ldots\}$ be a denumerable collection of (possibly infinite) subsets of $\Omega$. Then where $\check{p}$ is the disjunctive extension of $p$ and $\check{N}_j$ is as defined above

(1) If $\sum_{m=1}^{k} \check{p}(X^m)$ approaches a limit as $k$ increases without bound, then there are maximal $T$-consistent $\Phi_0$ and $\Phi_1$, subsets of $\Omega$, such that

$$\check{N}_0(\chi) \leqslant \lim_{k\to\infty} \sum_{m=1}^{k} \check{p}(X^m) \leqslant \check{N}_1(\chi).$$

(2) If $\sum_{m=1}^{k} \check{p}(X^m)$ increases without bound as $k$ increases without bound, then for every finite quantity $n$, there is some maximal $T$-consistent subset $\Phi_j$ of $\Omega$ such that $\check{N}_j(\chi) > n$.

*Proof*: The argument for clause (1) divides naturally in two parts.

(1a) That there is a $\Phi_0$ as stated in clause (1): For every $k \sum_{m=1}^{k} \check{p}(X^m) \leqslant \sum_{m=1}^{\infty} \check{p}(X^m)$ and so, by the corollary to lemma C, for every $k$ there is some $\Phi_j$ such that

$$\check{N}_j(\chi^k) \leqslant \sum_{m=1}^{\infty} \check{p}(X^m).$$

Thus by lemma D there is some $\Phi_j$ such that

$$\check{N}_j(\chi) \leqslant \sum_{m=1}^{\infty} \check{p}(X^m)$$

which establishes (1a).

(1b) That there is a $\Phi_1$ as stated in clause (1). Assume not. Then for every $\Phi_j$, $\check{N}_j(\chi) < \sum_{m=1}^{\infty} \check{p}(X^m)$ and, since this sum is finite, so is $\max_j \check{N}_j(\chi)$. Hence

$$\max_j \check{N}_j(\chi) < \sum_{m=1}^{\infty} \check{p}(X^m)$$

so, for some $k$

$$\max_j \check{N}_j(\chi) < \sum_{m=1}^{k} \check{p}(X^m).$$

By the remark (E), for some $k$

$$\max_j \check{N}_j(\chi^k) < \sum_{m=1}^{k} \check{p}(X^m).$$

In contradiction of the corollary to the preceding theorem. Thus (1b) and clause (1) are proved.

Clause 2. Assume the sum to increase without bound as $k$ increases. Then, for each $n$, there is some $k$ such that

$$n < \check{N}_j(\chi^k)$$

and hence, since for every maximal $T$-consistent $\Phi_j$ and every $k$, $\check{N}_j(\chi^k) \leqslant \check{N}_j(\chi)$, for each $n$ there is some $\Phi_j$ such that

$$n < \check{N}_j(\chi)$$

which establishes clause (2) and the theorem.

## VI.3 The Generalized Sum Condition Entails the Laws of Probability

We have seen that the laws of probability entail the generalized sum conditions under the assumptions that the logic in question is absolutely consistent, at least tautological, and compact. It was also established in IV.2 that under suitable constraints on $T$, any numerical function on sentences in $\Omega$ which satisfies the simple sum condition is a $T$-probability on $\Omega$. In the present section the discussion of the sum conditions is concluded in the demonstration that the generalized conditions are also sufficient for the laws of probability as applied to sets of sentences. The main theorem establishes that for suitable $T$, any numerical function which satisfies the conjunctive sum condition on subsets of $\Omega$ is the conjunctive extension of a $T$-probability on $\Omega$. The disjunctive extension of that probability may then be defined in the obvious way.

The route to this theorem is, if not obvious, at least not devious, and it commences, as might be expected, with the development of some consequences of the conjunctive sum condition.

The first theorem displays a simple relation between functions which satisfy the conjunctive sum condition and implication.

THEOREM 3.1. Let $\Omega$ be a denumerable field of sentences, $T$ be absolutely consistent and at least tautological, and $X$ and $Y$ be any subsets of $\Omega$. Then if no function which satisfies the conjunctive sum condition assigns a greater to value to $X$ than to $Y$, $X$ $T$-implies $Y$.

*Proof*: Assume that $X$ does not $T$-imply $Y$. Then for some maximal $T$-consistent $\Phi_j$, $X \subseteq \Phi_j$ and $Y \nsubseteq \Phi_j$. Now define the function $f$ on subsets of $\Omega$.

$$f(Z) = 1 \quad \text{if} \quad Z \in \Phi_j$$
$$f(Z) = 0 \quad \text{if} \quad Z \notin \Phi_j.$$

Then $f(X)=1>0=f(Y)$, and clearly $f$ satisfies the conjunctive sum condition.

We have now a theorem which develops the elementary probability properties.

THEOREM 3.2. For each of the following clauses let $\Omega$ be a denumerable field of sentences, $T$ be absolutely consistent and at least tautological, and assume that $\hat{p}$ and $\check{p}$ are functions defined on the subsets of $\Omega$ which satisfy the generalized sum conditions.

(1) $\qquad \hat{p}(\Lambda) = \check{p}(\Omega) = 1.$

*Proof:* Since for every $\Phi_j$, $\Lambda \subseteq \Phi_j$, and $\Phi_j \cap \Omega \neq \Lambda$ we have that

$$\hat{N}_j\{\Lambda\} = \check{N}_j\{\Omega\} = 1 \quad \text{for every } j.$$

Hence

$$\min_j N_j\{\Lambda\} = \max_j \hat{N}_j\{\Lambda\} = \min_j \check{N}_j\{\Omega\} = \max_j \check{N}_j\{\Omega\}$$

which, by the sum conditions, yields (1).

(2) $\qquad \check{p}(\Lambda) = \hat{p}(\Omega) = 0.$

*Proof:* By an argument dual to the preceding, since

$$\hat{N}_j\{\Omega\} = \check{N}_j\{\Lambda\} = 0 \quad \text{for every } j.$$

(3) $\qquad 0 \leqslant \hat{p}(X) \leqslant 1; \qquad 0 \leqslant \hat{p}(X) \leqslant 1.$

*Proof:* Directly from the sum conditions, since $0 \leqslant \hat{N}_j\{X\} \leqslant 1$ and $0 \leqslant \check{N}_j\{X\} \leqslant 1$ for every maximal $T$-consistent $\Phi_j$.

(4) If (every member of) $X$ is $T$-necessary, then $\hat{p}(X)=1$.

*Proof*: The hypothesis implies that $X$ is a subset of every maximal $T$-consistent $\Phi_j$.

(5) If $X$ intersects every maximal $T$-consistent $\Phi_j$, then $\check{p}(X)=1$.

(6) If $X$ is $T$-inconsistent then $\hat{p}(X)=0$.

(7) If $X$ intersects no maximal $T$-consistent $\Phi_j$, then $\check{p}(X)=0$.

(8) If $X$ intersects every maximal $T$-consistent $\Phi_j$ and the members of $X$ are pairwise $T$-incompatible, then

$$\check{p}(X) = \sum_{A \in X} p(A) = 1.$$

*Proof*: The hypothesis entails that for every $\Phi_j$, some $A \in X$ is a member of $\Phi_j$. Thus for every $\Phi_j$

$$\check{N}_j \{\{A\} | A \in X\} \geqslant 1$$

and the members of $X$ are pairwise $T$-incompatible, so no $\Phi_j$ includes more than one of them. Thus, for each $\Phi_j$,

$$\check{N}_j \{\{A\} | A \in X\} = 1.$$

From which (8) follows by the disjunctive sum condition and (5) above.

Given these basic properties, we turn now to a closer examination of the monotonicity and transparency properties of $\hat{p}$.

THEOREM 3.3. Let $\Omega$ be a denumerable field of sentences, $T$ be absolutely consistent and at least tautological, and assume that $\hat{p}$ satisfies the conjunctive sum condition. Then if $X$ and $Y$ are subsets of $\Omega$ and $X$ $T$-implies $Y$, $\hat{p}(X) \leqslant \hat{p}(Y)$. (Monotonicity of $\hat{p}$ for $T$-implication.)

*Proof*: The argument involves the application of the conjunctive sum condition to a certain collection of subsets of $\Omega$ which we now describe. Let $X = \{A_1, \ldots\}$. For each $n > 0$ let

$$Z_n = \{A_1, A_2, \ldots, A_{n-1}, -A_n\}.$$

So $Z_1 = \{-A_1\}$, $Z_2 = \{A_1, -A_2\}$, etc. and let $Z_0 = X$.

Let $\chi$ be the collection of all $T$-consistent $Z_i$ for $i > 1$. Since $X$ is denumerable, $\chi$ is at most denumerable, and is null if $X$ is $T$-necessary. In any event $\{X\} \cup \chi$ is not null.

Now if $\Phi_j$ is any maximal $T$-consistent subset of $\Omega$, then exactly one member of $\{X\} \cup \chi$ is a subset of $\Phi_j$, for if $\{-A_1\} \nsubseteq \Phi_j$, then $\{A_1\} \subseteq \Phi_j$, and either $\{A_1, -A_2\} \subseteq \Phi_j$ or $\{A_1, A_2\} \subseteq \Phi_j$, and so on. Further, the members of $\{X\} \cup \chi$ are pairwise $T$-incompatible.

Thus for each $\Phi_j$,

$$\hat{N}_j(\{X\} \cup \chi) = 1.$$

Also, in view of the fact that $X$ $T$-implies $Y$, if no member of $\chi$ is a subset of $\Phi_j$ then both $X$ and $Y$ are subsets of $\Phi_j$, hence, for each $\Phi_j$

$$\hat{N}_j(\{Y\} \cup \chi) \geqslant 1.$$

Thus, by the conjunctive sum condition

$$\hat{p}(X) + \sum_i \hat{p}(Z_i) = 1$$

$$\hat{p}(Y) + \sum_i \hat{p}(Z_i) \geqslant 1$$

so, $\hat{p}(X) \leqslant p(Y)$.

Under the assumptions of the theorem we have

COROLLARY 1. $\hat{p}$ is transparent for $T$-equivalence.

COROLLARY 2. If $X \subseteq Y$ then $\hat{p}(Y) \leqslant \hat{p}(X)$.

COROLLARY 3. If $X$ is finite then $\hat{p}(X) = p(\wedge X)$.

The remaining important characteristic of $\hat{p}$ to establish consequent upon the sum condition is the relationship of the values assigned to infinite sets to those assigned to finite sets.

THEOREM 3.4. Let $\Omega$, $T$, and $\hat{p}$ be as in the previous theorem. Then if $X \subseteq \Omega$ and $X_i$ ranges over the finite subsets of $X$:

$$\hat{p}(X) = \inf \{\hat{p}(X_i)\}.$$

*Proof*: By the preceding theorem, $\hat{p}(X) \leqslant \hat{p}(X_i)$ for each finite $X_i$. Thus there is a lower bound on the values of $\hat{p}$ for the various $X_i$ and hence there is a greatest lower bound, which can be no less than $\hat{p}(X)$.

$$\hat{p}(X) \leqslant \inf \{\hat{p}(X_i)\}.$$

So what remains to be shown is that for any positive $\varepsilon$ there is some finite $X_i$ such that

$$\hat{p}(X_i) - \hat{p}(X) < \varepsilon.$$

Let $X = \{A_1, A_2, \ldots\}$, and consider the covering nest of $X$

$$\{\{A_1\}, \{A_1, A_2\}, \{A_1, A_2, A_3\}, \ldots\}.$$

In view of the monotonicity of $p$ we may restrict $X_i$ to range over members of this nest, for any finite subset of $X$ is a subset of some member of the nest. We now employ a generalization of the construction used in the previous theorem. As there, let

$$Z_n = \{A_1, \ldots, A_{n-1}, -A_n\}$$

for $n \geqslant 1$, and for each $i$ let

$$\chi_i = \{Z_n | n \leqslant i\}.$$

So, for example,

$$\chi_3 = \{\{-A_1\}, \{A_1, -A_2\}, \{A_1, A_2, -A_3\}\}$$

and let

$$\chi = \bigcup_{i=1}^{\infty} \{\chi_i\}.$$

For convenience of reference we abbreviate

$$\sum_{n=1}^{i} \hat{p}(Z_n) \quad \text{as} \quad x_i$$

and

$$\sum_{n=1}^{\infty} \hat{p}(Z_n) \quad \text{as} \quad x.$$

Then

(i) $\qquad x = \sup \{x_i\}.$

(ii) For each $\Phi_j \in M(\Omega)$, exactly one member of $\chi \cup \{X\}$ is a subset of $\Phi_j$.

(iii) For each finite member $X_i$ of the above described covering nest of $X$, and each $\Phi_j \in M(\Omega)$, exactly one member of $\chi_i \cup \{X_i\}$ is a subset of $\Phi_j$;

(ii) and (iii) in company with the assumption that $\hat{p}$ satisfies the conjunctive sum condition entail that for each $X_i$

$$\hat{p}(X_i) + x = 1 = \hat{p}(X_i) + x_i.$$

So, by (i), for any positive $\varepsilon$, $i$ can be so selected that

$$\hat{p}(X_i) - \hat{p}(X) = x - x_i < \varepsilon$$

which proves the theorem.

The main theorem of this section is

THEOREM 3.5. Let $\Omega$ be a denumerable field of sentences, $T$ be absolutely consistent and at least tautological. Under these conditions if $\hat{p}$ is any numerical function on subsets of $\Omega$ which satisfies the conjunctive sum condition, and if $p$ is defined for the individual sentences of $\Omega$

$$p(A) = \hat{p}\{A\}.$$

Then (1) $\langle \Omega, T, p \rangle$ is a probability system and $\hat{p}$ is the conjunctive extension of $p$, and (2) The function $\check{p}$, defined on subsets of $\Omega$

$$\check{p}(X) = 1 - \hat{p}\{-A | A \in X\}$$

satisfies the disjunctive sum condition and is the disjunctive extension of $p$.

The proof of the first clause of this theorem is now virtually complete. That of the second clause involves yet another lemma.

*Proof of clause* (1). Clearly $p$ satisfies the simple sum condition of IV.3 and thus, by theorem IV.2.1, $\langle \Omega, T, p \rangle$ is a probability system. By corollary (3) to theorem 3.3, and the definition of the function $p$, $\hat{p}(X) = \mathrm{p}(\wedge X)$ for finite subsets of $\Omega$, and hence, by theorem 3.4, $\hat{p}(X) = \inf \{p(\wedge X_i)\}$ where $X_i$ ranges over the finite subsets of $X$. Thus $\hat{p}$ is the conjunctive extension of $p$.

For the proof of clause (2) we establish first a

*Lemma.* Under the assumptions of the theorem, if $\check{p}$ is defined on subsets of

$$\check{p}(X) = 1 - \hat{p}\{-A | A \in X\}$$

then $\check{p}(X) = p(\vee X)$ for finite subsets $X$ of $\Omega$, and

$$\check{p}(X) = \sup \{\check{p}(X_i)\}$$

where $X_i$ ranges over the finite subsets of $X$.

*Proof*: By clause (1) of the theorem, $\langle \Omega, T, p \rangle$ is a probability system, so $p$ is transparent for $T$-equivalence. Thus, for finite $X$

$$p(\vee X) = p(-\wedge\{-A | A \in X\}) = 1 - p(\wedge\{-A | A \in X\})$$

so, by clause (1) $= 1 - \hat{p}\{-A | A \in X\} = \check{p}(X)$.

Now let $X$ be denumerably infinite. Then, employing clause (1),

$$\begin{aligned} \check{p}(X) &= 1 - \hat{p}\{-A | A \in X\} = 1 - \inf\{\hat{p}\{-A | A \in X_i\}\} \\ &= \sup\{1 - \hat{p}\{-A | A \in X_i\}\} \\ &= \sup\{\check{p}(X_i)\} \end{aligned}$$

which completes the lemma.

It now follows that $\check{p}$ is the disjunctive extension of $p$, and by theorem 2.2, $\check{p}$ satisfies the disjunctive sum condition.

This completes the proof of the theorem.

# APPENDIX ON SET THEORY AND BOOLEAN ALGEBRAS

The reader is assumed to be familiar with the basic truths of naive set theory. Any determinate collection of objects is a *set*. If the object $a$ is a *member* of the set $A$, we write

$$a \in A.$$

The set $A$ is a *subset* of the set $B$ if every member of $A$ is also a member of $B$. In symbols;

$$A \subseteq B \Leftrightarrow \forall x \cdot x \in A \Rightarrow x \in B$$

A set is completely determined by its members, so two sets differ just in case some member of one of them is not a member of the other. Thus

$$A = B \Leftrightarrow A \subseteq B \quad \text{and} \quad B \subseteq A.$$

If every member of $A$ is a member of $B$, but $A \neq B$, then $A$ is said to be a *proper* subset of $B$.

If $S(x)$ is any condition which expresses a characteristic of objects, such as, for example,

$x$ is an even integer between one and nine

then we may write

$$\{x \mid S(x)\}$$

or

$$\{x \mid x \text{ is an even integer between one and nine}\}$$

to name the set of those objects each of which has that characteristic. In the case of the example, this is the set consisting of 2, 4, 6, and 8. This set may also be named by enclosing the names of its members in brackets

$$\{2, 4, 6, 8\} = \{x \mid x \text{ is an even integer between 1 and 9}\}.$$

If a set is infinite, or finite but very large, then it will be impossible or impractical to name it by enclosing the names of it members in brackets, and it will have to be named as the set of these objects satisfying a given condition. For example

$$\{x \mid x \text{ is a positive even integer}\}.$$

If the condition is as simple and obvious as this, then it may be assumed to be intuited from a small sequence of its instances, thus

$$\{2, 4, 6, \ldots\}$$

or, in the case of a large finite set

$$\{2, 4, 6, \ldots, 1{,}000\}.$$

Since sets are determined by their members, there is just one set which has no members, *the null set*, which is named

$$\Lambda.$$

If the condition $S(x)$ is impossible to satisfy, then

$$\Lambda = \{x \mid S(x)\}$$

so, for example

$$\Lambda = \{x \mid x \neq x\}.$$

The *union* of two sets is the set consisting of those objects which are members of either or both. The union of $A$ and $B$ is named

$$A \cup B$$

so

$$x \in A \cup B \Leftrightarrow x \in A \quad \text{or} \quad x \in B$$

and

$$A \cup B = \{x \mid x \in A \quad \text{or} \quad x \in B\}.$$

The *intersection* of two sets is the set consisting of those objects which are members of both sets. The intersection of $A$ and $B$ is named

$$A \cap B$$

so

$$x \in A \cap B \Leftrightarrow x \in A \quad \text{and} \quad x \in B$$

and

$$A \cap B = \{x \mid x \in A \quad \text{and} \quad x \in B\}.$$

The *difference* of $A$ and $B$, or the *relative complement* of $A$ and $B$, is the set of those objects which are members of $A$ but not of $B$. This set is named

$$A - B$$

so

$$x \in A - B \Leftrightarrow x \in A \quad \text{and} \quad x \notin B$$

and

$$A - B = \{x \mid x \in A \quad \text{and} \quad x \notin B\}.$$

We have, for all sets $A$, $B$, and $C$

$$A \cap B \subseteq A \subseteq A \cup B$$
$$\Lambda \subseteq A$$
$$A \cap \Lambda = \Lambda$$
$$A \cup \Lambda = A$$
$$(A - B) \cap (A - C) = A - (B \cup C)$$
$$(A - B) \cup (A - C) = A - (B \cap C)$$
$$(A - B) \cup (A \cap B) = A$$
$$(A - B) \cap (B - A) = \Lambda$$
$$(A \cap B) \cup (A - B) \cup (B - A) = A \cup B.$$

If discourse is restricted to subsets of a given set $Z$, then we may write simply

$$-A$$

for

$$Z - A$$

and in this case

$$-A \cap -B = -(A \cup B)$$
$$-A \cup -B = -(A \cap B)$$
$$A \subseteq B \Leftrightarrow -B \subseteq -A$$
$$A \cap -A = \Lambda$$
$$A \cup -A = Z$$
$$--A = A$$
$$A - B = A \cap -B.$$

Sets themselves are objects, and may themselves be members of other sets. Thus we may have, for example,

$$\begin{aligned} A &= \{2, 4, 6, 8\} \\ B &= \{1, 2, 3\} \\ \mathfrak{A} &= \{A, B\} \\ \mathfrak{B} &= \{\Lambda, A\}. \end{aligned}$$

In which case

$$\mathfrak{A} \cup \mathfrak{B} = \{A, B, \Lambda\}.$$

Since the null set is an object, if $C$ is any non-null set

$$\{\Lambda, C\} \neq \{C\}$$

although $C \cup \Lambda = C$ for each set $C$.

If $\mathfrak{C}$ is a set of sets, then the union of the members of $\mathfrak{C}$ is written

$$\bigcup \mathfrak{C}$$

so, referring to the above examples

$$\begin{aligned} &\bigcup \mathfrak{A} = \bigcup\{A, B\} = A \cup B = \{1, 2, 3, 4, 6, 8\} \\ &\bigcup \mathfrak{B} = \{\Lambda, A\} = \Lambda \cup A = A = \{2, 4, 6, 8\} \\ &\bigcup (\mathfrak{A} \cup \mathfrak{B}) = \bigcup \{A, B, \Lambda\} = A \cup B \cup \Lambda = A \cup B. \end{aligned}$$

If discourse is restricted to subsets of a given set $Z$, then for any set $\mathfrak{C}$ of sets

$$\bigcup \{-A \mid A \in \mathfrak{C}\} = -\bigcap \mathfrak{C}.$$

A set, the members of which are themselves sets, is commonly called a *collection*.

The notion of *order* is not inherent in that of set:

$$\{1, 2\} = \{2, 1\} = \{2, 2, 1\}.$$

It is helpful also to make use of ordered *sequences* of objects, which are distinguished in terms of their order as well as of their objects. We write

$$\langle a, b, c\rangle$$

for example, for the ordered triple consisting of $a$, $b$ and $c$ in that order, and similarly for ordered $n$-tuples of any finite length $n$. In distinction to the above example with sets

$$\langle 1, 2\rangle \neq \langle 2, 1\rangle$$

and both of these couples must be distinguished from the triples

$$\langle 2, 2, 1\rangle, \langle 2, 1, 2\rangle, \langle 1, 2, 2\rangle$$

which are also distinct from each other. The general law governing the identity of $n$-tuples is

$$\langle a_1, \ldots, a_n\rangle = \langle b_1, \ldots, b_n\rangle \Leftrightarrow$$
$$a_1 = b_1 \quad \text{and} \quad a_2 = b_2 \quad \text{and} \quad \ldots \quad \text{and} \quad a_n = b_n$$

or, more succinctly

$$\langle a_1, \ldots, a_n\rangle = \langle b_1, \ldots, b_n\rangle \Leftrightarrow \underset{1 \leqslant i \leqslant n}{\forall_i} \quad a_i = b_i.$$

We may also make use of denumerably infinite sequences. If $a$ is any sequence of length no greater than $j$ then $a(j)$ is the $j$'th member of $a$.

A finite set has just $n$ members for some non-negative integer $n$. If $A$ is finite and non-null then the members of $A$ correspond one-to-one with some set

$$\{x \mid 1 \leqslant x \leqslant n\}$$

and this set of positive integers may be used to *index* the members of $A$ thus

$$A = \{a_1, a_2, \ldots, a_n\}$$

or

$$A = \{a_i \mid 1 \leqslant i \leqslant n\}.$$

A set is *denumerably infinite* if it corresponds one-to-one with the set of all positive integers and is, accordingly, indexed by that set. Thus

$$A = \{a_1, a_2, \ldots\}$$
$$A = \{a_i \mid 1 \leqslant i\}$$

or, in this case, we may write

$$A = \{a_i\}$$

If a set is denumerably infinite then it may be put in one-to-one correspondence with some of its proper subsets. Thus, to pick a paradigm, the set of all positive integers corresponds one-to-one with the set of positive even integers. This characteristic distinguishes the denumerably infinite sets from the finite sets and has been proposed as a defining character of infinite sets.

The term *denumerable* is customarily applied to finite as well as to denumerably infinite sets.

Every subset of a denumerable set is denumerable. This follows from the preceding remark and the fact that every subset of the positive integers has a least element.

The union of finitely many denumerable sets is denumerable: This is easy to see by induction: If each of $A_1, \ldots, A_{n-1}, A_n$ is denumerable and $A_1 \cup A_2 \cup \ldots \cup A_{n-1}$ is also denumerable, then put the members of $A_1 \cup \ldots \cup A_{n-1}$ in one-to-one correspondence with the odd positive integers, and put the members of $A_n$ in one-to-one correspondence with the even positive integers. Then $A_1 \cup \ldots \cup A_n$ corresponds one-to-one with the set of positive integers and is hence denumerable.

Further, the union of denumerably many denumerable sets is also denumerable: Let

$$\mathcal{A} = \{A_1, A_2, \ldots\}$$

be a denumerable collection of denumerable sets. Let

$$\begin{aligned} A_1 &= \{a_1^1, a_1^2, a_1^3, \ldots\} \\ &\vdots \\ A_j &= \{a_j^1, a_j^2, a_j^3, \ldots\} \end{aligned}$$

etc. Then the problem of enumerating $\bigcup \mathcal{A}$ reduces to the problem of enumerating the pairs of positive integers. This may be done as follows: Let the pairs of integers be ordered according to the size of their sums, and within pairs which have the same sum, order them according to the magnitude of the first member.

$$\langle 1, 1\rangle, \langle 1, 2\rangle, \langle 2, 1\rangle, \langle 1, 3\rangle, \langle 2, 2\rangle, \langle 3, 1\rangle, \langle 1, 4\rangle \ldots.$$

There are infinite sets which are not denumerable. The following proposition provides a simple and famous example.

1.1. The set $K$ of all denumerably infinite sequences of zeros and ones is not itself denumerable.

*Proof:* Assume that $K$ is denumerable. It is clearly non-null, and not finite, so, on this assumption, it may be indexed by the positive integers, and

$$K = \{a_1, a_2, \ldots\}.$$

Each member $a_i$ of $K$ is a sequence

$$a_i = \langle a_i(1), a_i(2), a_i(3), \dots, a_i(j), \dots \rangle$$

and for each $i$ and each $j$, $a_i(j)$ is either zero or one.

Now consider the *diagonal sequence*, $b$, defined

$$\begin{aligned} b_1 &= 1 - a_1(1) \\ b_2 &= 1 - a_2(2) \\ &\vdots \\ b_j &= 1 - a_j(j) \\ &\vdots \end{aligned}$$

The sequence $b = \langle b_1, b_2, \dots \rangle$ differs from $a_1$ at the first place, from $a_2$ at the second place, and, for each $i$, $b$ differs from $a_i$ at the $i$'th place. Thus $b$ is distinct from each member of $K$ and is hence not a member of $K$. But $b$ is a denumerably infinite sequence of zeros and ones. The rejection of this absurdity forces us to reject the assumption that $K$ is denumerable.

An easy and important corollary to (1.1) is

1.2. If $A$ is a denumerably infinite set, then the set of all subsets of $A$ is infinite and not denumerable.

*Proof:* Let $A = \{a_1, a_2, \dots\}$. If $B$ is any subset of $A$, then $B$ corresponds to the sequence $b$, defined

$$\begin{aligned} b_i &= 1 \quad \text{if} \quad a_i \in B \\ b_i &= 0 \quad \text{if} \quad a_i \notin B. \end{aligned}$$

(So the null set corresponds to the sequence of all zeros, $A$ itself to the sequence of all ones, and so on.) Clearly the subsets of $A$ and the denumerable sequences of zeros and ones are in one-to-one correspondence, and hence if either is denumerable the other must be as well.

A further important distinction between finitude and denumerable infinity is revealed in the following

1.3. The set of all finite subsets of a denumerably infinite set is denumerably infinite.

*Proof:* Consider the set

$$\{0, 1, 01, 11, 001, 101, 110, 111, 0001, \dots\}$$

of all finite sequences of zeros and ones, none of which, save the first, ends in a zero. Clearly this set is denumerably infinite. (It is, in fact, just the non-negative integers in binary notation.) Now let

$$A = \{a_1, a_2, \ldots\}$$

be a denumerably infinite set, and let $B$ be any non-null finite subset of $A$. Since $B$ is finite, for some finite $n$, $a_n \in B$, but for all $m > n$, $a_m \notin B$. Thus $B$ corresponds to a sequence

$$b = \langle b_1, b_2, \ldots b_n \rangle$$

defined

$$\begin{aligned} b_i &= 1 \quad \text{if} \quad a_i \in B \\ b_i &= 0 \quad \text{if} \quad a_i \notin B. \end{aligned}$$

The null set corresponds to zero. This correspondence is evidently one-to-one, and thus the collection of finite subsets of $A$ is denumerable.

## A. BOOLEAN ALGEBRAS

Let $Z$ be a non-null collection of objects. A collection of subsets of $Z$ is a *Boolean algebra* of subsets of $Z$ if it satisfies the following conditions

(i) $Z \in \mathfrak{B}$.
(ii) If $A \in \mathfrak{B}$ and $B \in \mathfrak{B}$ then $A \cup B \in \mathfrak{B}$.
(iii) If $A \in \mathfrak{B}$ and $B \in \mathfrak{B}$ then $A - B \in \mathfrak{B}$.

As consequences of these characteristics we have also

(iv) $\Lambda \in \mathfrak{B}$.
(v) If $A \in \mathfrak{B}$ and $B \in \mathfrak{B}$ then $A \cap B \in \mathfrak{B}$.
(vi) If $\mathfrak{C}$ is a finite subcollection of $\mathfrak{B}$ then $\bigcup \mathfrak{C} \in \mathfrak{B}$.

*Proof* of (vi). Let $\mathfrak{C} = \{A_1, \ldots, A_n\}$. If $n = 1$ then $\bigcup \mathfrak{C} = A_1 \in \mathfrak{B}$. Now let $n$ be greater than 1 and assume that (vi) holds for subcollections with less members than $\mathfrak{C}$. Then $\bigcup \{A_1, \ldots A_{n-1}\} \in \mathfrak{B}$.

And, by (ii), $\bigcup \mathfrak{C} = \bigcup \{A_1, \ldots, A_{n-1}\} \cup A_n \in \mathfrak{B}$. Thus, (vi) follows from (ii) by mathematical induction on the size of $\mathfrak{C}$.

(vii) If $\mathfrak{C}$ is a finite subcollection of $\mathfrak{B}$ then $\bigcap \mathfrak{C} \in \mathfrak{B}$.

*Proof* of (vii). Writing $-A$ for $Z - A$, we have

$$\bigcap \mathfrak{C} = -\bigcup \{-A \mid A \in \mathfrak{C}\}.$$

By (vi), $\bigcup \{-A \mid A \in \mathfrak{C}\} \in \mathfrak{B}$, and by (iii) $\bigcap \mathfrak{C} \in \mathfrak{B}$. The principle (iii) of the definition of Boolean algebras may be replaced by

(iiia) If $A \in \mathfrak{B}$ then $Z - A \in \mathfrak{B}$,

to see this, we may argue that (iii) and (iiia) are equivalent in the presence of (i) and (ii). Clearly (ii) and (iii) imply (iiia). Now assume (iiia), and let $A$ and $B$ be any members of $\mathfrak{B}$. Then by (iiia), $Z-A \in \mathfrak{B}$, so by (ii), $(Z-A) \cup B \in \mathfrak{B}$. Or, more succinctly, since only subsets of $Z$ are under consideration;

$$-A \cup B \in \mathfrak{B}$$

so by (iiia), $-(-A \cup B) = A \cap -B = A - B \in \mathfrak{B}$.

## B. SOME EXAMPLES

Readers who are not familiar with Boolean Algebras may find it profitable to develop proofs of these examples. Examples (f) and (g) may be ignored in the absence of some understanding of recursion theory (not much is needed) since there is no essential reference to that theory in the body of the work. The method of argument is quite simple; just refer to the definition of Boolean algebras, or to its modification by adoption of (iiia) in place of (iii), and show whether or not the collection in question satisfies it.

(a) If $Z$ is any non-null set, then $\{\Lambda, Z\}$ is a Boolean algebra of subsets of $Z$.

(b) So $\{\Lambda, \{\Lambda\}\}$ is a Boolean algebra of subsets of $\{\Lambda\}$.

(c) If $Z$ is any non-null set, then the set of all subsets of $Z$ is a Boolean algebra of subsets of $Z$.

Let $I$ be the set $\{0, 1, 2, \ldots\}$ of non-negative integers. Then

(d) The set of all finite subsets of $I$ is *not* a Boolean algebra of subsets of I.

(e) The set $\mathfrak{B} = \{A \mid A \subseteq I$ and ($A$ is finite or $I - A$ is finite)$\}$ is a Boolean algebra of subsets of $I$.

(f) The set of all recursively decidable (i.e. recursive) subsets of $I$ is a Boolean algebra of subsets of $I$.

(g) The set of all recursively enumerable subsets of $I$ is *not* a Boolean algebra of subsets of $I$.

(h) The set of all sets of even non-negative integers is a Boolean algebra of subsets of the set of all even non-negative integers (by (c), above) but is *not* a Boolean algebra of subsets of $I$.

(i) Let $Z$ be the set of all denumerably infinite sequences of zeros and ones. Then $Z$ is non-denumerably infinite. For each integer $i \geqslant 1$, let $N_i$ be the set of those sequences in $Z$ each of which has a one at the $i$'th place. Then for each $i \geqslant 1$

$$\{N_i, (Z - N_i), \Lambda, Z\}$$

is a Boolean algebra of subsets of $Z$.

Let $x$ range over the members of $Z$, and for each $i$ let $x_i$ be the $i$'th member of $x$. So, for each $i$,

$$N_i = \{x \mid x_i = 1\}.$$

Thus, $N_1$ is the set of all sequences which begin with a one,

$$N_1 \cap -N_2$$

is the set of all sequences which begin 10..., and so on. We are in particular interested in picking out sets of sequences in terms of their initial segments. This is easily done; for each $n \geqslant 1$

$$\bigcap \{N_1, \ldots, N_n\}$$

is just the set of those sequences which begin with a segment of $n$ ones.

$$\bigcap \{-N_1, N_2, N_3, \ldots, N_n\}$$

is the set of those sequences which begin with a zero followed by $n-1$ ones, and so on. There are just $2^n$ different initial segments of length $n$. Any set that is formed from $\{N_1, \ldots, N_n\}$ by complementing some of its members will, when intersected, give one of these initial segments. For example, for $n = 3$ we have

$$\begin{aligned} \bigcap \{N_1, N_2, N_3\} \quad &= \{x \mid x_1 = 1, x_2 = 1, x_3 = 1\} \\ &\vdots \\ \bigcap \{-N_1, -N_2, -N_3\} &= \{x \mid x_1 = 0, x_2 = 0, x_3 = 0\}. \end{aligned}$$

To state this a bit more generally: For each positive integer $n$, a *constitution* of $\{N_1, N_2, \ldots, N_n\}$ is a set which includes for each $N_i \in \{N_1, N_2, \ldots, N_n\}$ just one of $N_i$, $-N_i$. So $\{N_1, N_2, \ldots, N_n\}$ is a constitution of itself, and for each $n \geqslant 1$ there are exactly $2^n$ constitutions of $\{N_1, N_2, \ldots, N_n\}$. An *n-segment* is the intersection of some constitution of $\{N_1, N_2, \ldots, N_n\}$. So each $n$-segment is a subset of $Z$, all the members of which have the same first $n$-elements. There are thus just $2^n$ distinct $n$-segments for each $n \geqslant 1$, which we denote

$$P^n_1, P^n_2, \ldots, P^n_{2^n}.$$

We have that;

(a) For each $n$ and $i \leqslant n$, $P^n_i \neq \Lambda$.

*Proof:* By induction on $n$. One may perhaps as clearly see intuitively that every finite sequence of zeros and ones begins some (indeed, infinitely many) sequences in $Z$.

(b) For each $n$, each member of $Z$ belongs to exactly one $n$-segment. (For each member of $Z$ begins in just one way.)

If $Z$ is any non-null set then a *partition* of $Z$ is a set $\mathfrak{P}$ of subsets of $Z$ such that

(a) $\Lambda \notin \mathfrak{P}$.
(b) If $A$ and $B$ are distinct members of $\mathscr{P}$, then $A \cap B = \Lambda$.
(c) $\bigcup \mathscr{P} = Z$.

So a partition is just a partitioning of $Z$ into pairwise disjoint and non-null subsets. In example (i) above, for each $n \geqslant 1$, the set $\{P^n_1, \ldots, P^n_{2^n}\}$ of $n$-segments is a partition of the set $Z$ of sequences.

A member $A$ of a Boolean algebra $\mathfrak{B}$ is said to be an *atom* of $\mathfrak{B}$ if

(i) $A \neq \Lambda$.
(ii) $\Lambda$ is the only member of $\mathfrak{B}$ which is a proper subset of $A$.

2.1. If $\mathfrak{B}$ is a finite Boolean algebra of subsets of the set $Z$, and $A_1, \ldots, A_k$ are all the atoms of $\mathfrak{B}$, then

$$\mathfrak{A} = \{A_1, \ldots, A_k\}$$

is a partition of $Z$.

*Proof:* Since the $A_i$ are atoms, they are non-null, so $\mathfrak{A}$ satisfies the first of the requirements for partitions. To see that $\mathfrak{A}$ also satisfies the second requirement, let $A_i$ and $A_j$ be distinct members of $\mathfrak{A}$. Then $A_i \cap A_j$ must be distinct from at least one of $A_i$, $A_j$, and is hence a proper subset of at least one of them. Thus, by the second characteristic of atoms, $A_i \cap A_j = \Lambda$.

Finally, to establish that $\mathfrak{A}$ is a partition, it will suffice to show that $\bigcup \mathfrak{A} = Z$. Since $\mathfrak{A} \subseteq \mathfrak{B}$, $\bigcup \mathfrak{A} \subseteq Z$ so it remains to show that $Z \subseteq \bigcup \mathfrak{A}$. We argue that if this is not so, then $\mathfrak{B}$ is not finite.

Assume that $Z$ is not a subset of $\bigcup \mathcal{A}$. Then $\bigcup \mathcal{A}$ is a proper subset of $Z$ and

$$Z - \bigcup \mathcal{A} \neq \Lambda.$$

Call this set, $Z - \bigcup \mathcal{A}$, $B_0$. Then

(a) $B_0$ is a member of $\mathcal{B}$ but is not an atom of $\mathcal{B}$.
(b) No atom is a subset of $B_0$.
(c) Some member of $\mathcal{B}$, call it $B_1$, is a proper subset of $B_0$.

Indeed, $B_1$ also satisfies the conditions (a), (b) and (c), so there is also a $B_2$, distinct from $B_0$ and $B_1$, which satisfies them, and so on. In this way we can define (inductively) a continuing sequence $B_0 \supseteq B_1 \supseteq B_2 \supseteq \ldots$. Each member of this sequence is a proper subset of all of its predecessors. Hence for any finite $n$ there are more than $n$ distinct members of $\mathcal{B}$. Thus the assumption that $Z$ is not a subset of $\bigcup \mathcal{A}$ is inconsistent with the assumption that $\mathcal{B}$ is finite. This establishes that $Z \subseteq \bigcup \mathcal{A}$ and we have that $\mathcal{A}$ is a partition of $Z$.

(2.1) shows that (and how) the atoms of a Boolean algebra form a partition. It is consistent with (2.1) that there should be finite Boolean algebras without atoms. In fact there are no such.

2.2. Every finite Boolean algebra has atoms.

*Proof:* Assume not. Let $\mathcal{B}$ be a finite Boolean algebra which has no atoms. Then every non-null member of $\mathcal{B}$ has a non-null proper subset which is a member of $\mathcal{B}$. Thus, as in the closing argument of the preceding proposition, there is sequence of distinct members of $\mathcal{B}$ of arbitrary length. Thus $\mathcal{B}$ is not finite.

2.3. If $\mathcal{B}$ is a finite Boolean algebra, then every non-null member of $\mathcal{B}$ is the union of atoms of $\mathcal{B}$.

*Proof:* Given $\Lambda \neq A \in \mathcal{B}$, let $\mathcal{A}(A)$ be the set of all those atoms of $\mathcal{B}$ each of which intersects $A$. Then if $B \in \mathcal{A}(A)$, $B \cap A \neq \Lambda$ and hence, since $B \cap A \subseteq B$, $B \cap A = B$, so $B \subseteq A$. Thus $\mathcal{A}(A)$ consists of just those atoms of $\mathcal{B}$ which are subsets of $A$.

Clearly, then, $\bigcup \mathcal{A} \subseteq A$.

Now, suppose that some element $x$ of $Z$ is not a member of $\bigcup \mathcal{A}$. Since the atoms of $\mathcal{B}$ partition $Z$, $x$ is a member of just one atom $C$ of $\mathcal{B}$. $C$ is not a subset of $\bigcup \mathcal{A}$ and hence not a member of $\mathcal{A}$. Thus $C$ does not intersect $A$, so $x \notin A$. Thus $A \subseteq \bigcup \mathcal{A}$.

A finite Boolean algebra always has atoms which form a partition of its domain. The converse relation also holds.

2.4. If $Z$ is any non-null set and $\mathfrak{A}$ is any finite partition of $Z$, then there is just one Boolean algebra of which the members of $\mathfrak{A}$ are the atoms. This algebra is finite.

*Proof:* Given $\mathfrak{A}$ let $\mathfrak{B}$ be the set consisting of $\Lambda$ and in addition all finite unions of members of $\mathfrak{A}$. Then

(i) $\mathfrak{B} \neq \Lambda.$

(ii) If $A$ and $B$ are members of $\mathfrak{B}$, then both are finite unions of members of $\mathfrak{A}$, hence $A \cup B$ is also such a union, and $A \cup B \in \mathfrak{B}$.

(iii) If $A$ and $B$ are members of $\mathfrak{B}$, then let $\mathfrak{C}$ consist of those members of $\mathfrak{A}$ which are subsets of $A$ but not of $B$. Then $\bigcup \mathfrak{C} = A - B$, so $A - B$ is also a finite union of members of $\mathfrak{A}$ and is hence a member of $\mathfrak{B}$.

(iv) $Z = \bigcup \mathfrak{A} \in \mathfrak{B}.$

(i)–(iv) are just the defining characteristics of Boolean algebras. Thus the set $\mathfrak{B}$ is a Boolean algebra.

To see that the atoms of an algebra determine it uniquely, assume $\mathfrak{B}_1$ and $\mathfrak{B}_2$ to be distinct finite Boolean algebras of subsets of the set $Z$, and assume that some subset $A$ of $Z$ is a member of $\mathfrak{B}_1$ but not of $\mathfrak{B}_2$. Then by (2.3), $A$ is the union of atoms of $\mathfrak{B}_1$. Thus some atom of $\mathfrak{B}_1$ is not an atom of $\mathfrak{B}_2$.

Any collection of subsets of a given non-null set $Z$ forms a unique Boolean algebra in either of two equivalent ways. One way is to build up the closure of the collection under the Boolean operations of finite union and complement with respect to $Z$. That is to say, to define the algebra to consist of all finite unions and relative complements of members of the given collection, of all finite unions and relative complements of members of the thus augmented collection, and so on. The other way is to take the intersection of all those Boolean algebras of which the given collection is a subset. One needs to see that these are equivalent, and that the result is always a Boolean algebra of subsets of $Z$.

Let $Z$ be a non-null set and let $\mathfrak{A}$ be a collection of subsets of $Z$. Let $\mathfrak{A}^*$ be the result of closing $\mathfrak{A}$ under the operations of finite union and complement with respect to $Z$. Then clearly $\mathfrak{A}^*$ is a Boolean algebra of subsets of $Z$, and $\mathfrak{A} \subseteq \mathfrak{A}^*$.

Now let $\mathfrak{B}$ be the intersection of all Boolean algebras which include $\mathfrak{A}$ as a subset. $\mathfrak{A}^*$ is such an algebra, so $\mathfrak{B} \subseteq \mathfrak{A}^*$. Further, if $\mathfrak{B}'$ is any Boolean algebra which includes $\mathfrak{A}$, it must also include the closure of $\mathfrak{A}$ under finite unions and complements. Thus $\mathfrak{A}^*$ is a subset of every such algebra, and is hence also a subset of their intersection. Thus $\mathfrak{A}^* \subseteq \mathfrak{B}$, and $\mathfrak{A}^* = \mathfrak{B}$.

In view of this relation, we may speak of *the* Boolean algebra of subsets of *Z generated by the collection* $\mathfrak{A}$ *of subsets of Z.*

If $\mathfrak{A}$ is a finite partition of $Z$, then the Boolean algebra of subsets of $Z$ generated by $\mathfrak{A}$ is the Boolean algebra of which the members of $\mathfrak{A}$ are the atoms.

Example (i) continued. For each $n$, the collection

$$\mathfrak{L}_n = \{P_1^n, \ldots, P_{2^n}^n\}$$

of $n$-segments is a partition of $Z$. So (a) these $2^n$ $n$-segments are the atoms of the finite Boolean algebra $\mathfrak{B}(\mathfrak{L}_n)$ of subsets of $Z$. Every non-null member of $\mathfrak{B}(\mathfrak{L}_n)$ is a union of some or all of the $n$-segments $P_1^n, \ldots, P_{2^n}^n$. Thus (b) If $x \in Z$, then $\{x\}$ is a member of no $\mathfrak{B}(\mathfrak{L}_n)$. (Show that every $P_i^n$, and hence every non-null member of $\mathfrak{B}(\mathfrak{L}_n)$ is a non-denumerably infinite subset of $Z$).

The Boolean algebra $\mathfrak{B}'$ is said to be an *extension* of the Boolean algebra $\mathfrak{B}$ if both are Boolean algebras of subsets of the same set $Z$ and $\mathfrak{B} \subseteq \mathfrak{B}'$.

(c) (Example (i) continued.) If $m \geqslant n$, then $\mathfrak{B}(\mathfrak{L}_m)$ is an extension of $\mathfrak{B}(\mathfrak{L}_n)$. (Show that every atom of $\mathfrak{B}(\mathfrak{L}_n)$ is a member of $\mathfrak{B}(\mathfrak{L}_m)$.)

(d) Let $\mathfrak{B} = \bigcup_{n=1}^{\infty} \{\mathfrak{B}(\mathfrak{L}_n)\}$. Then $\mathfrak{B}$ is a Boolean algebra.

*Proof:* Clearly $Z$ and $\Lambda$ are members of $\mathfrak{B}$, so it remains to show only that $\mathfrak{B}$ is closed under relative complement and the formation of unions. The first of these is pretty obvious. The second depends upon (c) above.

(e) If $x \in Z$, then $\{x\} \notin \mathfrak{B}$. (Refer to (b) and to the definition of $\mathfrak{B}$.)

## C. SIGMA-ALGEBRAS

Let $Z$ be an infinite set. If $\mathfrak{A}$ is a finite collection of subsets of $Z$, then $\bigcup \mathfrak{A}$ and $\bigcap \mathfrak{A}$ are well defined; an element of $Z$ is an element of $\bigcup \mathfrak{A}$ just in case it is an element of some member of $\mathfrak{A}$, and the elements of $\bigcap \mathfrak{A}$ are just those objects which are elements of every member of $\mathfrak{A}$. Our comprehension of unions and intersections apparently does not depend upon the collections which are united or intersected being finite. We may understand

$$x \in \bigcup \mathfrak{A} \Leftrightarrow \text{ for some } A \in \mathfrak{A}, x \in A$$
$$x \in \bigcap \mathfrak{A} \Leftrightarrow \text{ for every } A \in \mathfrak{A}, x \in A$$

when $\mathcal{A}$ includes infinitely many subsets of a given set $Z$ as well as when $\mathcal{A}$ is finite. (Here, as always, we must be careful to distinguish the cardinality of $\mathcal{A}$ from the cardinalities of its members.)

Some idea of the distinction between finite (as in Boolean algebras) and denumerable unions may be gained from examples: The set

$$\mathcal{B} = \{A \mid A \subseteq I \text{ and } (A \text{ is finite or } I - A \text{ is finite})\}$$

is a Boolean algebra (Cp. example (e), p.168). But there are denumerably infinite subsets of $\mathcal{B}$, the unions of which are not members of $\mathcal{B}$. For example, the set

$$E = \{0, 2, 4, 6, 8, \ldots\}$$

of non-negative even integers is an infinite subset of $I$, and its complement, the set of non-negative odd integers, is also infinite. Thus $E$ is not a member of $\mathcal{B}$. But each of the sets

$$\{0\}, \{2\}, \{4\}, \ldots$$

is a member of $\mathcal{B}$, so the set

$$\{\{0\}, \{2\}, \{4\}, \ldots\}$$

is a subset of $\mathcal{B}$, and the union of this set

$$\bigcup \{\{0\}, \{2\}, \{4\}, \ldots\}$$

is just $E$. Thus

3.1. If $\mathcal{B}$ is a Boolean algebra and $\mathcal{A}$ is a finite subset of $\mathcal{B}$, then $\bigcup \mathcal{A} \in \mathcal{B}$, but this is not in general true for denumerably infinite subsets $\mathcal{A}$ of Boolean algebras. That is to say, there are Boolean algebras $\mathcal{B}$, such that for some denumerably infinite subset $\mathcal{A}$ of $\mathcal{B}$, $\bigcup \mathcal{A} \notin \mathcal{B}$.

Or, more briefly, we say that Boolean algebras are closed under the formation of finite unions, but not in general closed under the formation of denumerably infinite unions.

The force of *not in general* must be remarked here. Some Boolean algebras are closed under the formation of denumerable unions. A finite Boolean algebra provides a trivial example of such. Consider, also, the collection of all sets of non-negative integers. It is a Boolean algebra of subsets of $I$ (Cp. example (c) p.168). If $\mathcal{A}$ is any denumerable collection of sets of non-negative integers, then $\bigcup \mathcal{A}$ is also a set of non-negative integers, and so the collection of all sets of non-negative integers is a

Boolean algebra which is closed under the formation of denumerable unions.

Another instancc of lack of closure under the formation of denumerable unions, accessible only through the rudiments of recursion theory and avoidable without prejudice to understanding the text, is provided by the Boolean algebra of the set of all recursive sets of non-negative integers. There are non-recursive sets of integers. Let $A$ be such. $A$ is denumerable, and for each integer $n \in A$, the set $\{n\}$ is recursive. Thus the set

$$\bigcup \{\{n\} \mid n \in A\} = A$$

is, though not recursive, the union of a denumerable collection of recursive sets.

Some Boolean algebras are closed under the formation of denumerable unions and some are not. Those that are are of particular interest for probability theory and for logic. They are called *sigma algebras*, and are defined by strengthening the definition of Boolean algebras.

Let $Z$ be a non-null set. A collection $\mathscr{S}$ of subsets of $Z$ is a *sigma algebra* of subsets of $Z$ if $\mathscr{S}$ satisfies the following conditions.

(i) $Z \in \mathscr{S}$.
(ii) If $\mathscr{A}$ is a denumerable subcollection of $\mathscr{S}$, then $\bigcup \mathscr{A} \in \mathscr{S}$.
(iii) If $A$ and $B$ are members of $\mathscr{S}$, then $A - B$ is also a member of $\mathscr{S}$.

As in the definition of Boolean algebras, (iii) is equivalent to

(iiia) If $A \in \mathscr{S}$ then $Z - A \in \mathscr{S}$

in the presence of (i) and (ii). To see this it need only be remarked that the second clause of the above definition entails the corresponding clause in the definition of Boolean algebras. This also assures that sigma-algebras are Boolean algebras.

Sigma-algebras are also closed under the formation of denumerable intersections. Let $\mathscr{S}$ be a sigma-algebra and $\mathscr{A}$ a denumerable subcollection of $\mathscr{S}$. Then, by (iiia), for each $A \in \mathscr{A}$ $-A \in \mathscr{S}$. So, by (ii)

$$\bigcup \{-A \mid A \in \mathscr{A}\} \in \mathscr{S}$$

and by (iiia) $\bigcap \mathscr{A} = -\bigcup \{-A \mid A \in \mathscr{A}\} \in \mathscr{S}$.

To each Boolean algebra there corresponds in an obvious and important way a sigma algebra which includes it. To reveal this some auxiliary notions will be helpful.

An *upward nest* of sets is a sequence $A_1 \subseteq A_2 \subseteq \dots$ each member of which is a subset of all of its successors. A *downward nest* is a sequence $A_1 \supseteq A_2 \supseteq \dots$ each member of which is a superset of all of its successors. The *limit* of an upward nest is just the union of its members. This set includes every member of the nest as a subset, and is a subset of any set which shares this characteristic. The *limit* of a downward nest is a subset of each member of the nest, and is a superset of every other set which shares this characteristic. The limit of a downward nest is the intersection of it members. It should be remarked that this notion of limit is set-theoretical, and must be distinguished from the numerical concept. The relation between the two is developed in the next appendix.

A collection of sets is closed under the formation of limits if it includes the limit of every denumerable nest formed from its members. A collection of sets which has this property is said to be *monotone*.

Not all Boolean algebras are monotone, as our earlier examples reveal. Of course the limit of every *finite* nest of members of a Boolean algebra is always a member of the algebra, but there may be a *denumerable* sequence of members of a Boolean algebra which forms a nest, the limit of which is not a member of the algebra. The above reference to example (e) may be improved again here: The set of all subsets of $I$ which are either finite or have a finite complement is a Boolean algebra. The sequence

$$\{0\}, \{0, 2\}, \{0, 2, 4\}, \dots$$

is an upward nest of elements of this algebra. The limit of this sequence – the set of even non-negative integers – is however not a member of the algebra.

Monotonicity is exactly the mark of a sigma-algebra.

3.2. Every sigma-algebra is monotone. Every monotone Boolean algebra is a sigma-algebra.

*Proof:* Sigma-algebras are closed under denumerable unions and intersections. Thus the limit of every denumerable nest of members of the algebra is a member of the algebra.

If $\mathcal{B}$ is a Boolean algebra which is monotone, then if $\mathcal{A}$ is any denumerable subcollection of $\mathcal{B}$

$$\mathcal{A} = \{A_1, A_2, \dots\}$$

then the sequence

$$A_1, A_1 \cup A_2, A_1 \cup A_2 \cup A_3, \dots$$

is an upward nest of members of $\mathfrak{B}$. Thus, since $\mathfrak{B}$ is monotone, the limit of this sequence, which is just $\bigcup \mathfrak{A}$, is a member of $\mathfrak{B}$. Thus $\mathfrak{B}$ is closed under the formation of denumerable unions, and in view of the other clauses of the definitions, is a sigma-algebra.

Here are some more useful theorems about monotone collections.

3.3. If a Boolean algebra contains the limit of each of its denumerable upward nests it is monotone. If a Boolean algebra contains the limit of each of its denumerable downward nests it is monotone.

*Proof:* Assume that the Boolean algebra $\mathfrak{B}$ of subsets of $Z$ contains the limit of each of its denumerable upward nests, and let

$$\mathfrak{A} = \{A_1 \supseteq A_2 \supseteq \dots\}$$

be a denumerable downward nest of members of $\mathfrak{B}$. Then

$$\mathfrak{A}' = \{-A_1 \subseteq -A_2 \subseteq \dots\}$$

is an upward nest of members of $\mathfrak{B}$, so $\bigcup \{-A_i\} \in \mathfrak{B}$ and $-\bigcup \{-A_i\} = \bigcap \{A_i\} = \bigcap \mathfrak{A} \in \mathfrak{B}$.

The proof of the second clause is completely symmetrical.

3.4. If $\Gamma$ is a collection of monotone classes, then $\bigcap \Gamma$ is monotone.

*Proof:* If $A_1, A_2, \dots$ is an upward (downward) nest of members of $\bigcap \Gamma$, then it is also an upward (resp. downward) nest of members of every member of $\Gamma$. Thus the limit of this nest is a member of every member of $\Gamma$, and is hence a member of $\bigcap \Gamma$.

3.5. If $Z \neq \Lambda$ then the collection of all subsets of $Z$ is monotone.

3.6. If $\mathfrak{A}$ is any non-null collection of subsets of the non-null set $Z$, then $\mathfrak{A}$ has a monotone extension, and the intersection of the collection of all monotone extensions of $\mathfrak{A}$ is not null.

*Proof:* Directly from (3.4) and (3.5).

If $\mathfrak{B}$ is a Boolean algebra of subsets of the set $Z$, then a *Boolean extension* of $\mathfrak{B}$ is a Boolean algebra of subsets of $Z$ of which $\mathfrak{B}$ is a subcollection.

3.7. If $\mathfrak{B}$ is a Boolean algebra of subsets of $Z$ and $\Gamma$ is any non-null collection of Boolean extensions of $\mathfrak{B}$, then $\bigcap \Gamma$ is a Boolean algebra.

*Proof:* Directly from the definition of Boolean algebras.

3.8. Let $\mathfrak{B}$ be any Boolean algebra of subsets of the set $Z$, let $\Gamma$ be the collection of all monotone Boolean extensions of $\mathfrak{B}$, and let $S(\mathfrak{B})=\bigcap \Gamma$. Then $S(\mathfrak{B})$ is not null, is monotone, is a Boolean algebra, and is a subset of every monotone Boolean extension of $\mathfrak{B}$.

*Proof:* The collection of all subsets of $Z$ is a monotone Boolean extension of $\mathfrak{B}$, so $\Gamma \neq \Lambda$. $\mathfrak{B}$ itself is a subset of every member of $\Gamma$, so $\bigcap \Gamma = S(\mathfrak{B}) \neq \Lambda$. (3.4) establishes the monotonicity of $S(\mathfrak{B})$, and (3.7) entails that it is Boolean. Since every Boolean extension of $\mathfrak{B}$ is a member of $\Gamma$, $\bigcap \Gamma = S(\mathfrak{B})$ is a subset of every such collection.

3.9. $S(\mathfrak{B})$ is a sigma-algebra and is a subset of every sigma-algebra which is an extension of $\mathfrak{B}$.

*Proof:* By (3.2) and (3.8) $S(\mathfrak{B})$ is a sigma-algebra. Now let $\mathcal{S}$ be any sigma-algebra which is an extension of $\mathfrak{B}$. Then by (3.2) $\mathcal{S}$ is monotone, and is, of course, a Boolean algebra. Thus by (3.8), $S(\mathfrak{B}) \subseteq \mathcal{S}$.

The development culminating in (3.9) is important for our understanding of the foundations of probability and the theory of belief, in particular for the question of the nature of our beliefs involving infinities. We can see pretty clearly in that theorem and its support how any denumerable Boolean algebra extends in a natural and plausible way to a unique 'smallest' sigma-algebra, a sigma-algebra which both includes the original Boolean algebra and is also a sub-collection of every sigma-algebra which does this. In the sequel we continue to refer to this extension of a Boolean algebra $\mathfrak{B}$ as $S(\mathfrak{B})$.

## D. CONSTITUTIONS AND INDEPENDENCE

If $\mathfrak{A}$ is any denumerable collection of subsets of a given set $Z$, then a *constitution* of $\mathfrak{A}$ is any collection of sets including for each $A \in \mathfrak{A}$, either $A$ or $Z-A$ (not both) and including no other sets. Thus $\mathfrak{A}$ is a constitution of itself, the set of complements of members of $\mathfrak{A}$ is a constitution of $\mathfrak{A}$, and so on. If $\mathfrak{A}$ is of finite size $k$, then there are just $2^k$ distinct constitutions of $\mathfrak{A}$. If $\mathfrak{A}$ is denumerably infinite, then there are non-denumerably many distinct constitutions of $\mathfrak{A}$, for they correspond one-to-one with the subsets of $\mathfrak{A}$. These truths assume some importance in the theory of probability.

If $Z$ is non-null and $\mathfrak{A}$ is any finite, non-null collection of subsets of $Z$, then there is some constitution $\mathfrak{A}'$ of $\mathfrak{A}$ such that $\bigcap \mathfrak{A}' \neq \Lambda$. This may be seen by induction on the size of $\mathcal{A}$. Assume it true for all collections smaller than a given size $n \geqslant 1$, and let

$$\mathcal{A} = \{A_0, A_1, \ldots, A_{n-1}, A_n\}.$$

By the inductive assumption $\{A_0, \ldots, A_{n-1}\}$ has some constitution $\mathcal{B}$ such that $\bigcap \mathcal{B} \neq \Lambda$. If $\bigcap \mathcal{B} \cap A_n = \Lambda$ then $\bigcap \mathcal{B} \subseteq -A_n$, and $\bigcap \mathcal{B} \cap -A_n \neq \Lambda$. Thus not both $\bigcap \mathcal{B} \cap A_n$ and $\bigcap \mathcal{B} \cap -A_n$ are null, and since both

$$\mathcal{B} \cup \{A_n\} \quad \text{and} \quad \mathcal{B} \cup \{-A_n\}$$

are constitutions of $\mathcal{A}$, it follows that some constitution of $\mathcal{A}$ has a non-null intersection.

There are also collections of sets every constitution of which has a non-null intersection. Here is an example.

Let
$$\begin{aligned} Z &= \{1, 2, 3, 4, 5, 6, 7, 8\} \\ A &= \{1, 2, 3, 4\} \\ B &= \{1, 2, 5, 6\} \\ C &= \{1, 3, 5, 7\} \end{aligned}$$

Then it may, and should, be seen with a little computation that every constitution of $\{A, B, C\}$ has a non-null intersection.

Example (i) continued. The Boolean algebra $\mathcal{B} = \bigcup_{n=1}^{\infty} \{\mathcal{B}(P_n)\}$ is not a sigma algebra. (Show, by reference to (b) page 173, that the limit of the nest $\{N_1, N_1 \cap N_2, \ldots\}$ is not a member of $\mathcal{B}$, and hence that $\mathcal{B}$ is not monotone.)

If $Z \neq \Lambda$ and $\mathcal{A}$ is any collection of subsets of $Z$, then $\mathcal{A}$ generates a Boolean algebra $\mathcal{B}$ of subsets of $Z$, which in turn is extended uniquely to $S(\mathcal{B})$, the smallest sigma-algebra which includes $\mathcal{B}$. We may in some cases denote this sigma-algebra by $S(\mathcal{A}, Z)$, to indicate its origin in $\mathcal{A}$, or even by $S(\mathcal{A})$ when the context permits.

If $\mathcal{A}$ is a denumerable and independent collection of subsets of the non-null set $Z$, then the sigma algebra $S(\mathcal{A})$ is said to have $\mathcal{A}$ as a *basis*. From the point of view of probability theory, sigma-algebras with finite bases are of particular interest. Such algebras are finite, and are determined by their bases in an interesting and simple way.

4.1. Let $\mathcal{A}$ be a finite and independent collection of subsets of $Z$. Then if $A$ is any non-null member of $S(\mathcal{A}, Z)$, there are constitutions $\mathcal{A}_1, \mathcal{A}_2, \ldots, \mathcal{A}_k$ of $\mathcal{A}$ such that

$$A = \bigcap \mathcal{A}_1 \cup \bigcap \mathcal{A}_2 \cup \ldots \cup \bigcap \mathcal{A}_k.$$

The proof of 5.1, which is not quite simple, is omitted. A theorem quite like it, more directly relevant to the purposes of the work is developed in Chapter V.

Not every sigma-algebra has a basis. For example, let

$$Z = \{1, 2, 3\}$$
$$\mathfrak{B} = \{\Lambda, Z, \{1\}, \{2\}, \{3\}, \{1, 2\}, \{2, 3\}, \{1, 3\}\}$$

then there is no independent collection of sets from which $\mathfrak{B}$ is generated. This may, and should, be seen by inspecting the members of $\mathfrak{B}$.

It can be shown, an analogous theorem is established in Chapter V, that

5.2. A finite sigma-algebra has a basis if and only if it has exactly $2^k$ atoms for some finite integer $k$.

Example (i) continued. Each of the algebras $\mathfrak{B}(P_n)$ has a basis (by (a) p. 173). In each case $\{N_1, \ldots N_n\}$ (and any of its constitutions as well) is a basis of $\mathfrak{B}(P_n)$. To show this, show, by reference to 2.4, p. 172, that

$$\mathfrak{B}(P_n) = \mathfrak{B}(\{N_1, \ldots, N_n\})$$

for each $n$.

### E. QUOTIENT ALGEBRAS

If $\mathcal{S}$ is a sigma-algebra of subsets of $Z$ and $A$ is any non-null subset of $Z$ then the collection

$$\{A \cap B \mid B \in \mathcal{S}\}$$

is called the *quotient algebra* of $\mathcal{S}$ with $A$. It is usually named $\mathcal{S}/A$.

Some facts about quotient algebras: First

5.3. If $\mathcal{S}$ is a sigma-algebra of subsets of $Z$ and $\Lambda \neq A \subseteq Z$, then $\mathcal{S}/A$ is a sigma-algebra of subsets of $A$.

*Proof:* $A \cap Z = A$ so $A \in \mathcal{S}/A$, and the first condition of sigma-algebras is established.

Let $\mathfrak{A} = \{A_1, A_2, \ldots\}$ be a denumerable subcollection of $\mathcal{S}/A$. Then there is a denumerable subcollection $\mathfrak{B} = \{B_1, B_2, \ldots\}$ of $\mathcal{S}$ such that $A_i = A \cap B_i$, for each $i = 1, 2, \ldots$. Since $\mathcal{S}$ is a sigma-algebra, $\bigcup \mathfrak{B} \in \mathcal{S}$, so $A \cap \bigcup \mathfrak{B} = \bigcup \mathfrak{A} \in \mathcal{S}/A$, which establishes the second condition.

Finally, if $C = A \cap B \in \mathcal{S}/A$, where $B \in \mathcal{S}$, then $Z - B \in \mathcal{S}$ so $A \cap (Z - B) \in \mathcal{S}/A$, and – since $A \subseteq Z$ – $A - B \in \mathcal{S}/A$.

5.4. If $\mathcal{S}$ is a sigma-algebra of subsets of $Z$ then $\mathcal{S}/Z = \mathcal{S}$.

# APPENDIX ON MEASURE THEORY

If a one-square-acre plot of ground is partitioned into a finite number of sub-plots, then it is easy to think of the collection of all combinations of those sub-plots as a Boolean algebra of subsets of the set of points in the inclusive plot. That is to say, that if the set of points in the inclusive plot is $Z$, then the collection $\mathfrak{B}$ of all combinations of the given subplots of $Z$ has the characteristics

(i) $Z \in \mathfrak{B}$.
(ii) If $A$ and $B$ are members of $\mathfrak{B}$ then $A \cup B \in \mathfrak{B}$.
(iii) If $A \in \mathfrak{B}$ then $Z - A \in \mathfrak{B}$.

Also, if $\mathfrak{A}$ is the collection of subplots into which $Z$ is partitioned, then for each $B \in \mathfrak{B}$, there are $A_1, A_2, \ldots, A_k$ members of $\mathfrak{A}$ such that the $A_i$ are pairwise disjoint and

$$B = A_1 \cup A_2 \cup \ldots \cup A_k.$$

It is quite natural in this case to think of the *proportional area*, the proportion of $Z$ occupied, of the various plots in $\mathfrak{B}$ as given by a function $\mu$ which assigns some number between zero and one to each member of $\mathfrak{B}$. Such a function will be defined for each member of $\mathfrak{B}$ and will have the characteristics;

$$\mu(Z) = 1$$
$$\mu(\Lambda) = 0.$$

If $A \subseteq B$ then $\mu(A) \leqslant \mu(B)$.
If $A \cap B = \Lambda$ then $\mu(A \cup B) = \mu(A) + \mu(B)$

$$0 \leqslant \mu(A) \leqslant 1.$$

Something like these simple intuitions must be at the bottom of the mathematical theory of measures. What these intuitions and that theory have to do with probability is a topic discussed in the body of the text.

If $\mathfrak{B}$ is a Boolean algebra of subsets of a non-null set $Z$, then a *measure* on $\mathfrak{B}$ is a numerical function $\mu$ which assigns a non-negative number to each member of $\mathfrak{B}$ in such a way that if $A$ and $B$ are members of $\mathfrak{B}$ such that $A \cap B = \Lambda$, then

$$\mu(A \cup B) = \mu(A) + \mu(B).$$

A measure on a Boolean algebra $\mu$ of subsets of $Z$ is said to be *normal* if it assigns 1 to $Z$.

If $\mu$ is a normal measure on the Boolean algebra $\mathfrak{B}$ of subsets of $Z$, then for all $A$ and $B$ members of $\mathfrak{B}$

(1) $\mu(-A) = 1 - \mu(A).$

*Proof:* $\mu(A) + \mu(-A) = \mu(Z) = 1.$

Perhaps the simplest normal measures are those which assign only the values zero and one. In the case of finite Boolean algebras, in particular, they may be easily characterized:

Let $\mathfrak{B}$ be a finite Boolean algebra. Then a *two-valued measure* on $\mathfrak{B}$ is a function which assigns one to some atom of $\mathfrak{B}$ and to every superset of that atom, and assigns zero to every other member of $\mathfrak{B}$.

It is easily seen that the appellation 'two-valued measure' is appropriate, that is, that two-valued measures are normal measures according to the above definitions. It is also easily established that

(i) If $\mu$ is a normal measure on a finite Boolean algebra $\mathfrak{B}$ which assigns only the values zero and one, then $\mu$ is a two-valued measure on $\mathfrak{B}$.

*Proof:* If $\mu$ assigns zero to every atom then – since $Z$ is just the union of the atoms – $\mu(Z)=0$, so $\mu$ is not normal. Thus $\mu(A)=1$ for some atom $A$. If there are distinct atoms $A_1$ and $A_2$ to both of which $\mu$ assigns one, then $\mu(A_1 \cup A_2)=2$, again contradicting the assumption that $\mu$ is normal. Thus every normal measure which assigns only the values zero and one assigns one to some atom and zero to every other atom. By (2.3) p. 171, every superset of the selected atom is assigned one. If $A$ is the atom to which $\mu$ assigns one and $A$ is not a subset of the set $B$, then, since the atoms form a partition, $A \subseteq -B$ so $\mu(-B)=1$. So only supersets of the selected atom are assigned one. And, indeed;

(ii) If $\mu$ is a normal measure on a finite Boolean algebra $\mathfrak{B}$ which assigns to each member of $\mathfrak{B}$ one of two values, then $\mu$ is a two-valued measure. (For $\mu(\Lambda)=0$ and $\mu(Z)=1$.)

If $\mathfrak{B}$ is a finite Boolean algebra then there are just as many distinct two-valued measures on $\mathfrak{B}$ as there are atoms of $\mathfrak{B}$. If the atoms of $\mathfrak{B}$

are $A_1, \ldots, A_k$, then a *weighting* of the atoms of $\mathfrak{B}$ is a set

$$w = \{w_1, \ldots, w_k\}$$

of non-negative numbers which sum to one:

$$w_i \geqslant 0 \qquad \sum_i w_i = 1.$$

Each weighting of the atoms of a finite Boolean algebra $\mathfrak{B}$ determines a normal measure in an obvious way.

(iii) Let $\mathfrak{B}$ be a finite Boolean algebra and let $v_1, \ldots, v_\kappa$ be the two-valued measures which correspond to the atoms $A_1, \ldots, A_k$ of $\mathfrak{B}$. Let $w$ be any weighting of the atoms of $\mathfrak{B}$. Then the function $\mu$ defined for members $B$ of $\mathfrak{B}$

$$\mu(B) = \sum_{i=1}^{\kappa} \omega_i \cdot v_i(B)$$

is a normal measure on $\mathfrak{B}$.

*Proof:* Of course, since $\mu$ is the weighted average of quantities all of which are either zero or one, its values will never exceed one nor be exceeded by zero. Since $v_i(Z) = 1$ for each $i$,

$$\mu(Z) = \sum_{i=1}^{\kappa} w_i = 1.$$

To see the additivity of $\mu$ notice first that for each $B \in \mathfrak{B}$

$$\sum_{i=1}^{\kappa} v_i(B)$$

is just the number of atoms of $\mathfrak{B}$ which are subsets of $B$. Thus if $A$ and $B$ have a null intersection, so that no atom is a subset of both, then for each $i$, $1 \leqslant i \leqslant k$,

$$v_i(A \cup B) = v_i(A) + v_i(B).$$

So, for each $i$,

$$\begin{aligned} w_i \cdot v_i(A \cup B) &= w_i(v_i(A) + v_i(B)) \\ &= w_i \cdot v_i(A) + w_i \cdot v_i(B) \end{aligned}$$

and

$$\mu(A \cup B) = \sum_{i=1}^{\kappa} w_i \cdot v_i(A \cup B)$$
$$= \sum_{i=1}^{\kappa} w_i \cdot v_i(A) + \sum_{i=1}^{\kappa} w_i \cdot v_i(B)$$
$$= \mu(A) + \mu(B).$$

We have also the converse of (iii):

(iv) If $\mu$ is any normal measure on the finite Boolean algebra $\mathfrak{B}$ then there is a weighting $w$ of the atoms of $\mathfrak{B}$ such that for each $B \in \mathfrak{B}$

$$\mu(B) = \sum_{i=1}^{\kappa} w_i \cdot v_i(B)$$

where the $v_i$ are the two-valued measures which correspond to the atoms of $\mathfrak{B}$.

*Proof:* Just let $w(A) = \mu(A)$ for each atom $A$ of $\mathfrak{B}$. Then $w$ is a weighting of the atoms of $\mathfrak{B}$. Let $B$ be any member of $\mathfrak{B}$ and $A_1, \ldots, A_k$ the atoms of $\mathfrak{B}$. Assume for simplicity of illustration, and without loss of generality, that $A_1, \ldots, A_n$ are the atoms which are subsets of $B$. Then $B = A_1 \cup \ldots \cup A_n$ so $\mu(B) = \sum_{i=1}^{n} \mu(A_i)$. Since $v_i(B) = 1$ for each $i$ from $l$ to $n$

$$\mu(B) = \sum_{i=1}^{n} \mu(A_i) v_i(B)$$

now each of the atoms $A_{n+1}, \ldots, A_k$ is not a subset of $B$, so each of $v_{n+1}, \ldots, v_\kappa$ assigns zero to $B$. Hence

$$\mu(A_i) v_i(B) = 0$$

for each $i$, $n+1 \leqslant i \leqslant k$. Hence

$$\sum_{i=n+1}^{\kappa} \mu(A_i) \cdot v_i(B) = 0$$

and

$$\sum_{i=1}^{n} \mu(A_i) \cdot v_i(B) = \sum_{i=1}^{n} \mu(A_i) \cdot v_i(B) + \sum_{i=n+1}^{\kappa} \mu(A_i) v_i(B)$$

$$\sum_{i=1}^{n} \mu(A_i) \cdot v_i(B) = \sum_{i=1}^{\kappa} \mu(A_i) v(B)$$

which establishes (iv).

If $\mathfrak{B}$ is a finite Boolean algebra, $\mu_1, \ldots, \mu_\kappa$ are normal measures on $\mathfrak{B}$, and $v$ is a weighting of $\mu_1, \ldots, \mu_\kappa$ then the function

$$\mu = \sum_{i=1}^{\kappa} v_i \cdot \mu_i$$

is a normal measure on $\mathfrak{B}$.

(2) $0 \leqslant \mu(A) \leqslant 1.$

(3) $A \subseteq B \Rightarrow \mu(A) \leqslant \mu(B).$

*Proof:* $A \subseteq B \Rightarrow A \cap -B = \Lambda$

$$\Rightarrow \mu(A) + \mu(-B) = \mu(A \cup -B)$$
$$\Rightarrow \mu(A) + (1 - \mu(A \cup -B)) = \mu(B)$$
$$\Rightarrow \mu(A) \leqslant \mu(B).$$

(4) If $\mathfrak{A}$ is a finite subcollection of $\mathfrak{B}$, the members of which are pairwise disjoint, then

$$\mu(\bigcup \mathfrak{A}) = \sum_{A \in \mathfrak{A}} \mu(A).$$

*Proof:* By induction on the size of $\mathfrak{A}$, from the additivity of measures over disjoint pairs.

(5) $\mu(\Lambda) = 0.$

(6) $\mu(A \cup B) = \mu(A) + \mu(B) - \mu(A \cap \mathrm{B}).$

Measures on Boolean algebras are not without interest, but by far the more important development, from the point of view of probability theory as well as in general, involves measures defined on sigma-algebras. The distinction depends upon the distinction between *finite* additivity ((4) above) and *denumerable* additivity:

If $R = \{r_1, r_2, \ldots\}$ is a denumerably infinite set of non-negative real numbers, then the various sums

$$r_1, r_1 + r_2, r_1 + r_2 + r_3, \ldots$$

or, as they are more perspicuously written

$$\sum_{i=1}^{1} r_i, \sum_{i=1}^{2} r_i, \sum_{i=1}^{3} r_i, \ldots$$

are called *partial sums* of the set $R$. The sequence of these sums may be thought of as continuing indefinitely, as the sums are taken for more and more members of $R$. The sequence is also denumerable. Since the numbers

$r_i$ are all non-negative, later members of this sequence are always at least as large as their predecessors. That is to say, as the index $n$ increases through the positive integers, the sum

$$r_1 + r_2 + \cdots + r_n = \sum_{i=1}^{n} r_i$$

never decreases. (It is said to be *monotonically non-decreasing* as $n$ increases.) There are three possibilities as far as the manner of growth of partial sums is concerned.

(I) They increase without bound. That is to say, for any positive number $q$, there is some positive integer $n$, such that

$$q < \sum_{i=1}^{n} r_i.$$

This would be the case if the set $R$ were the positive integers, so that for each $i$, $r_i = i$.

(II) For some certain $n_0$, every succeeding member of the sequence is equal to

$$\sum_{i=1}^{n_0} r_i.$$

In this case, all the numbers in $R$ beyond $r_{n_0}$ are zero.

(III) The sums continue to increase, so for each integer $n$, there is some integer $m$ such that

$$\sum_{i=1}^{n} r_i < \sum_{i=1}^{m} r_i.$$

But there is some real number $q$, such that for every integer $n$

$$\sum_{i=1}^{n} r_i < q.$$

In this case there will be some least number $q_0$ with this property: No partial sum will be equal to $q_0$, but they will approach $q_0$ as a limit. That is to say, that for any positive quantity $\varepsilon$, however small, there will be some $n$ such that

$$(q_0 - \varepsilon) < \sum_{i=1}^{n} r_i < q_0.$$

This situation may be variously described as follows.

$$q_0 = \lim_{n \to \infty} \sum_{i=1}^{n} r_i$$

$$q_0 = \sup\left\{\sum_{i=1}^{n} r_i\right\}.$$

In this case the sequence of partial sums is said to *converge* to $q_0$, and the sum of all the members of $R$ is defined to be $q_0$:

$$\sum_{i=1}^{\infty} r_i = \sum_{r \in R} r = q_0.$$

Here is an example. Let $R$ consist of the fractions

$$\tfrac{1}{2}, \tfrac{1}{4}, \tfrac{1}{8}, \ldots$$

so for each $i$, $r_i = \frac{1}{2}^i$.

Then for each $n$

$$\sum_{i=1}^{n} r_i = \sum_{i=1}^{n} \tfrac{1}{2}^i = 2^n/2^{n+1}$$

so,

$$\sum_{i=1}^{\infty} r_i = \lim_{n \to \infty} \sum_{i=1}^{n} r_i = 1.$$

In the case (I), the sequence of partial sums increases without bound, and the sum $\sum_{i=1}^{\infty} r_i$ is left undefined. In the case (II) this infinite sum is equal to the finite sum $\sum_{i=1}^{no} r_i$.

As with partial sums, so with sequences or sets of numbers in general: If $R$ is a set of real numbers (which for our purposes may be assumed to be non-negative) then $R$ may or may not have a greatest member. If (I) $R$ has a greatest member, say $n$, then $n$ is of course the least upper bound of the members of $R$: No member of $R$ exceeds $n$, and any number which exceeds all members of $R$ must also exceed $n$. If $R$ has no greatest member, then either (II) $R$ has no upper bound; for any number $n$, there is some member $m$ of $R$ such that $m > n$, or (III) There is some number $n$ such that

(a) If $r \in R$ then $r < n$.

(b) If $m > r$, for every $r \in R$, then $n \leqslant m$. In this case (III), we shall have also

(c) If $\varepsilon$ is any positive quantity, then for some $r \in R$, $n-\in<r$. And the number $n$ is the least upper bound, or supremum, of the numbers in $R$.

In cases (I) and (III) we write

$$\sup R = n.$$

When consideration is restricted to non-negative numbers, the situation of lower bounds is not quite symmetrical to that of upper bounds, for every set of non-negative numbers has a lower bound, and thus a greatest lower bound. Thus the relevant distinction is between sets of numbers which include a least member and those which do not. In the first instance

(i) For some $n \in R$, $n$ exceeds no member of $R$.

and in the second

(ii) There is some $n$ such that
(a) For every $r \in R$, $n<r$.
(b) If $m<r$ for every $r \in R$, then $m \leqslant n$

and consequently

(c) If $\in$ is any positive quantity, then for some $r \in R$, $r<n+\in$.

In either case, $n$ is the infimum or greatest lower bound of $R$, and we write

$$\inf R = n.$$

In view of the definition of infinite sums under certain conditions, we can in some cases take the sum of the values of a measure $\mu$ for a denumerably infinite number of sets in a Boolean algebra, when that algebra is also a sigma-algebra. The condition under which an infinite sum is defined is that there be an upper bound on the sequence of partial sums, a number which is exceeded by no partial sum in the sequence.

Let $\mathfrak{B}$ be a Boolean algebra of subsets of $Z$, and let $S(\mathfrak{B})$ be the smallest sigma-algebra which is an extension of $\mathfrak{B}$. If $\mu$ is a normal measure defined for the members of $\mathfrak{B}$, there may be members of $S(\mathfrak{B})$ for which $\mu$ is not defined. We can thus raise the question of how appropriately to extend the definition of the measure $\mu$ to the sigma-algebra $S(\mathfrak{B})$. One obvious constraint suggests itself: It is that the measure $\mu$ should be additive over disjoint unions of members of $S(\mathfrak{B})$. If $\mathfrak{A}=\{A_1, A_2, \ldots\}$ is a denumerable collection of pairwise disjoint members of $S(\mathfrak{B})$, then, for each $n$

$$\sum_{i=1}^{n} \mu(A_i) = \mu\left(\bigcup_{i=1}^{n} \{A_i\}\right) \leqslant 1.$$

Thus the sum of a normal measure over a denumerable collection of pairwise disjoint sets will be defined, and we can consistently require that

(*DA*) If $\mathcal{A}$ is a denumerable subcollection of $S(\mathcal{B})$, the members of which are pairwise disjoint, then

$$\mu(\bigcup \mathcal{A}) = \sum_{A \in \mathcal{A}} \mu(A).$$

Here are some reasons in favor of this constraint.

(i) It entails finite additivity: If $\mu$ is denumerably additive on $S(\mathcal{B})$, then $\mu$ is finitely additive on $\mathcal{B}$.

(ii) It entails monotonicity of $\mu$ for the subset relation, no set is assigned a smaller number than any of its subsets. Thus it conforms to the vague but strong intuition that larger sets should be assigned larger values by $\mu$, at least in a case where that intuition is clear.

(iii) This reason is best stated as a theorem.

1.1. If $\mathcal{B}$ is a denumerable Boolean algebra, $S(\mathcal{B})$ is the smallest sigma-algebra which is an extension of $\mathcal{B}$, and $\mu$ is a normal measure which is defined on $\mathcal{B}$, then the extension $\mu'$ of $\mu$ to $S(\mathcal{B})$ is denumerably additive if and only if

$$\mu'(\lim \mathcal{A}) = \lim_{A \in \mathcal{A}} \mu'(A)$$

for every nest $\mathcal{A}$ of members of $S(\mathcal{B})$.

*Proof:* We argue first that every denumerably additive normal measure satisfies the equality stated in the consequent.

Let $\mathcal{A}$ be an upward nest in $S(\mathcal{B})$.

$$\mathcal{A} = \{A_1 \subseteq A_2 \subseteq \cdots \subseteq \cdots\}$$

and define the sequence $\{B_i\}$

$$\begin{cases} B_1 = A_1 \\ B_{i+1} = A_{i+1} - A_i \end{cases}$$

then the collection of sets $\{B_i\}$ is denumerable, and these sets are pairwise disjoint. Thus by (*DA*)

$$\mu'(\bigcup \{B_i\}) = \sum_{i=1}^{\infty} \mu'(B_i)$$

also

$$\bigcup \{B_i\} = \bigcup \{A_i\} = \lim \{A_i\}.$$

Finally, for each $n$

$$\sum_{i=1}^{n} \mu'(B_i) = \mu'(A_n).$$

(Proof of this last by an obvious induction.)

So

$$\lim_{n\to\infty} \sum_{i=1}^{n} \mu'(B_i) = \lim_{n\to\infty} \mu'(A_n)$$

and

$$\mu'(\lim \mathfrak{A}) = \lim_{A\in\mathfrak{A}} \mu'(A).$$

Thus the condition holds for upward nests.

Now let $\mathfrak{A}=\{A_1 \supseteq A_2 \supseteq \cdots \supseteq\}$ be a downward nest in $S(\mathfrak{B})$. Then $\mathfrak{A}'=\{-A_1 \subseteq -A_2 \subseteq \cdots \subseteq\}$ is an upward nest in $S(\mathfrak{B})$ and

$$\begin{aligned}\mu'(\lim \mathfrak{A}') &= \lim_{-A_i\in\mathfrak{A}'} \mu'(-A_i)\\ &= \lim_{A_i\in\mathfrak{A}} (1 - \mu'(A_i))\\ &= 1 - \lim_{A_i\in\mathfrak{A}} \mu'(A_i)\end{aligned}$$

so

$$\begin{aligned}\mu'(\bigcup \{-A_i\}) &= 1 - \lim_{A_i\in\mathfrak{A}} \mu'(A_i)\\ \mu'(-\bigcap \{A_i\}) &= 1 - \lim_{A_i\in\mathfrak{A}} \mu'(A_i)\\ 1 - \mu'(\bigcap \{A_i\}) &= 1 - \lim_{A_i\in\mathfrak{A}} \mu'(A_i)\\ \mu'(\lim \mathfrak{A}) &= \lim_{A_i\in\mathfrak{A}} \mu'(A_i).\end{aligned}$$

It now remains to show that the equality of limits entails the denumerable additivity of $\mu'$.

Let $\mathfrak{A}=\{A_i, A_2, \ldots\}$ be a denumerable collection of pairwise disjoint members of $S(\mathfrak{B})$. For each $i$ let

$$B_i = A_1 \cup A_2 \cup \cdots \cup A_i.$$

Then

$$\mathcal{A}' = \{B_1 \subseteq B_2 \subseteq \cdots \subseteq\}$$

is an upward nest in $S(\mathcal{B})$ so

$$\mu(\bigcup \mathcal{A}') = \lim_{i \to \infty} \mu'(B_i).$$

Now for each $n$,

$$\mu'(B_n) = \sum_{i=1}^{n} \mu'(A_i)$$

for the $A_i$ are pairwise disjoint, and $\mu'$ is finitely additive.

Further,

$$\bigcup \mathcal{A}' = \bigcup \mathcal{A}$$

so

$$\mu'(\bigcup \mathcal{A}) = \lim_{n \to \infty} \sum_{i=1}^{n} \mu'(A_i)$$
$$= \sum_{i=1}^{\infty} \mu'(A_i).$$

This theorem associates two important intuitions; that of denumerable additivity, and that which relates the measure of a limit set to the limit of the measures of its subsets. It shows that these are equivalent, and accordingly, supports the following definition of normal measure on an arbitrary sigma-algebra.

If $\mathcal{S}$ is a sigma-algebra of subsets of $Z$, then a *normal measure* on $\mathcal{S}$ is a function $\mu$ which assigns values to the members of $\mathcal{S}$ such that

(i) $0 \leqslant \mu(A) \leqslant 1$ for every member $A$ of $\mathcal{S}$.
(ii) $\mu(Z) = 1$.
(iii) If $\mathcal{A}$ is a denumerable collection of pairwise disjoint members of $\mathcal{S}$ then $\mu(\bigcup \mathcal{A}) = \sum_{A \in \mathcal{A}} \mu(A)$.

## A. RELATIVE MEASURES

If $\mathcal{S}$ is a sigma-algebra of subsets of $Z$ and $A \subseteq Z$, then the quotient algebra $\mathcal{S}/A$ is also a sigma-algebra. $\mathcal{S}/A$ is a sigma-algebra of subsets of $A$. A case of particular interest is that in which a normal measure $\mu$ is defined on a sigma-algebra, $\mathscr{S}$, and $A$ is some member of $\mathscr{S}$ to which $\mu$ assigns a positive value. If we think of the measure $\mu$ as assigning to each member of $Z$ its proportional area, the proportion of $Z$ which it occupies, then we may also think of raising an analogous question with respect to proportional areas within $A$. If $B$ is a member of $\mathscr{S}$, the proportional area of $A$ occupied by $B$ will be the proportion of the area of $A$ which is occupied by $A \cap B$. The precise development of this intuition flows from the following definition.

If $\mu$ is a normal measure defined on a sigma-algebra $\mathscr{S}$ of subsets of $Z$ and $A \in \mathscr{S}$ such that $\mu(A) \neq 0$, then the *relativization* of $\mu$ to $A$ is defined for each $B \in \mathscr{S}$

$$\mu_A(B) = \frac{\mu(A \cap B)}{\mu(A)}.$$

Here are some easy and illuminating consequences. In each case the assumptions of the definition are presupposed.

(2.1) $\mu_A(A) = 1$.

(2.2) If $\mathscr{B} = \{B_1, B_2, \ldots\}$ is a denumerable collection of pairwise disjoint members of $\mathscr{S}/A$ then

$$\mu_A(\bigcup \mathscr{B}) = \sum_{B_i \in \mathscr{B}} \mu_A(B_i).$$

*Proof:* Each $B_i \in \mathscr{B}$ is a member of $\mathscr{S}/A$ and is hence a subset of $A$. So $\bigcup \mathscr{B}$ is a subset of $A$ and

$$A \cap \bigcup \mathscr{B} = \bigcup \mathscr{B}.$$

Thus

$$\mu_A(\bigcup \mathscr{B}) = \frac{\mu(A \cap \bigcup \mathscr{B})}{\mu(A)} = \frac{\mu(\bigcup \mathscr{B})}{\mu(A)}.$$

Each $B_i$ is a subset of $A$ so

$$\mu_A(B_i) = \frac{\mu(A \cap B_i)}{\mu(A)} = \frac{\mu(B_i)}{\mu(A)}$$

for each $i$. Thus

$$\sum_{B_i \in \mathscr{B}} \mu_A(B_i) = \frac{\sum_{B_i \in \mathscr{B}} \mu(B_i)}{\mu(A)}.$$

Finally, the sets $B_i$ are also pairwise disjoint members of $\mathcal{S}$, and $\mu$ is a normal measure on $\mathcal{S}$. Thus

$$\mu(\bigcup \mathcal{B}) = \sum_{B_i \in \mathcal{B}} \mu B_i)$$

from which (2.2) follows. Thus $\mu_A$ is a normal measure on the members of $\mathcal{S}/A$.

(3.1) If $A \subseteq B$ then $\mu_A(B) = 1$.

(3.2) $\mu_A(Z) = 1$.

(3.3) If $B \in \mathcal{S}$ and $\mu(A \cap B) = 0$, then $\mu_A(B) = 0$.

(3.4) If $B \in \mathcal{S}$ and $A \cap B = \Lambda$ then $\mu_A(B) = 0$.

(3.5) If $B \in \mathcal{S}$ then $\mu_A(B) = \mu_A(A \cap B)$.

Normal measures on sigma-algebras are the basic devices of probability theory. Some illustration of this may be provided by the further development of an earlier example.

The set of all denumerably infinite sequences of zeros and ones is non-denumerably infinite. As in example (i) of p. 169 denote this set by $Z$, and, also as in that example, for each $i$ let $N_i$ be the set of those sequences each of which has a one at the $i$'th place. From the standpoint of probability theory, the set $Z$ is taken to represent all possible histories of sequential occurrence and non-occurrence of some chance event, such as, for example, that an ideal coin comes to rest heads up, in a denumerably infinite sequence of trials. If one represents occurrence and zero non-occurrence, then the set $N_i$ consists of just those sequences in which the event occurs at the $i$'th trial, and this set may be identified with that occurrence. Thus, in the example of the coin, $N_i$ is the event

heads at the $i$'th toss

and $Z - N_i$ is the complementary event

tails at the $i$'th toss.

The algebraic structure of composite events is now natural and facile:

$$N_i \cap N_j$$

is the occurrence of heads at both the $i$'th and the $j$'th toss, and

$$N_i \cup N_j$$

is the occurrence of heads at at least one of these.

$$\bigcap_i \{N_i\}$$

is the occurrence of heads at every trial and

$$\bigcup_i \{N_i\}$$

is the occurrence of heads at some trial.

The sigma-algebra of all subsets of $Z$ may be considered a sigma-algebra of events, under the assumptions that complements and denumerable unions of events are again events. $Z$ is the necessary event, $\Lambda$ the impossible event.

Our intuitions about *chance* conform pretty well to the constraints of normal measures on sigma-algebras. For simplicity of illustration, let us assume that the chance of heads on any trial is $\frac{1}{2}$. We may write this

$$\text{I} \qquad c(N_i) = \tfrac{1}{2}, \forall i$$

assuming $c$ to be a normal measure on the sigma-algebra $\mathcal{S}$ of subsets of $Z$. Then, as consequences of this assumption and the definition of normal measures

$$c(Z) = 1 = c(N_i) + c(-N_i)$$
$$c(-N_i) = \tfrac{1}{2}$$
$$c(N_i \cap N_j) + c(N_i \cap -N_j) = \tfrac{1}{2}$$

etc.

It is important to our understanding of chance that these constraints do not suffice completely to determine a normal measure $c$ on $\mathcal{S}$. To see this, consider a simple sub-algebra $\mathcal{S}_2$ of $\mathcal{S}$, consisting of just $N_1$, $N_2$, and the Boolean combinations of these. This algebra includes

$$\Lambda, N_1 \cap N_2, N_1 \cap -N_2, -N_1 \cap N_2, -N_1 \cap -N_2$$

and all unions of these. It is a finite Boolean algebra, and hence a sigma-algebra. Both $c$ and $c'$ are normal measures on $\mathcal{S}_2$

$$c(N_1 \cap N_2) = c(N_1 \cap -N_2) = c(-N_1 \cap N_2)$$
$$= c(-N_1 \cap -N_2) = \tfrac{1}{4}$$
$$c(N_1) = \tfrac{1}{2} = c(N_2)$$
$$c'(N_1 \cap N_2) = c'(-N_1 \cap -N_2) = \tfrac{5}{16}$$
$$c'(N_1 \cap -N_2) = c'(-N_1 \cap N_2) = \tfrac{3}{16}$$
$$c'(N_1) = \tfrac{1}{2} = c'(N_2)$$

and both $c$ and $c'$ assign $\frac{1}{2}$ to each of $N_1$, $N_2$.

To see the difference between $c$ and $c'$ it is helpful to employ the concept of *relative* or *conditional* chance. We may think not only of the chance of an event, but also of the chance of one event given that another has happened. In some cases, these may be the same, if, in particular, the events have nothing to do with each other. But in other cases they may differ significantly. An extreme example is the contrast between (a) the chance of an event, and (b) the chance of the same event given that it has happened. The second of these, if defined, must always be one. A less extreme contrast is that between (a) the chance of two fair coins both coming up heads, and (b) the chance of this given that the first coin has come up heads. The first of these, under standard assumptions, is one-fourth, while the second is just the chance that the second will come up heads as well, or one-half.

Relative chances are represented by relative measures. If $c$ gives the chances of events in the sigma-algebra $\mathcal{S}$, then, assuming $c(A)$ to be distinct from zero, $c_A$ gives the chances of events in $\mathcal{S}$ given that $A$ has happened. The plausibility of this may be supported, in considerable part, by regarding the characteristics (2.1)–(3.5) of relative measures in this light.

An event $A$ is said to be *independent* of an event $B$ if the chance of $B$ is the same as the chance of $B$ given $A$. That is to say, if $c(B)=c_A(B)$, assuming that $c_A$ is defined. Under this assumption, we have that

$$c(B) = c_A(B) \Leftrightarrow c(A)\cdot c(B) = c(A \cap B).$$

The right member of this equivalence may be taken as the definition of independent events, since it permits ignorance of the value of $c(A)$. The definition is formulated in terms of normal measures in general.

If $\mu$ is a normal measure on a sigma-algebra $\mathcal{S}$ then the members $A$ and $B$ of $\mathcal{S}$ are *independent in* $\mu$ if

$$\mu(A \cap B) = \mu(A)\cdot\mu(B).$$

Here are some consequences of this definition. In each case $A$ and $B$ are members of $\mathcal{S}$, $\mu$ is a normal measure on $\mathcal{S}$, and, whenever employed, relative measures are assumed to be defined.

(4.1) $A$ and $B$ are independent in $\mu \Leftrightarrow B$ and $A$ are independent in $\mu$.

(4.2) $A$ is independent of $Z$ in $\mu$, and $A$ is independent of $\Lambda$ in $\mu$.

(4.3) If $A$ is independent of $B$ in $\mu$ then $A$ is independent of $-B$ in $\mu$.

Now the difference between the measures $c$ and $c'$ on the sigma-algebra $\mathcal{S}_2$ may be clearly remarked: $N_1$ and $N_2$ are independent in $c$ but not in $c'$.

$$c_{N_1}(N_2) = c(N_2) = \tfrac{1}{2}$$
$$c'_{N_1}(N_2) = \tfrac{5}{8} \neq \tfrac{1}{2} = c'(N_2).$$

To return now to the measure $c$, which was intended to represent the notion of *chance* on the sigma-algebra $\mathcal{S}$ of all infinite sequences of zeros and ones. If we assume that the tosses of the coin represented by the members of these sequences are independent then we have

If $i \neq j$, then $N_i$ and $N_j$ are independent in $c$.

This constraint, however, added to those previously stipulated, is still not sufficient to assure that each $N_i$ is independent of the intersections of all other events. It is consistent with it that every pair $N_i$, $N_j$, for distinct $i$ and $j$, are independent, but that, for example, there are $N_i$, $N_j$, and $N_k$, with $i$, $j$, and $k$ all distinct, such that $N_i$ is not independent of $N_j \cap N_k$. To see this, consider the finite sigma-algebra $\mathcal{S}_3$, consisting of all the Boolean functions of $N_1$, $N_2$, and $N_3$. And consider a measure $d$ which apportions chances as indicated in this diagram.

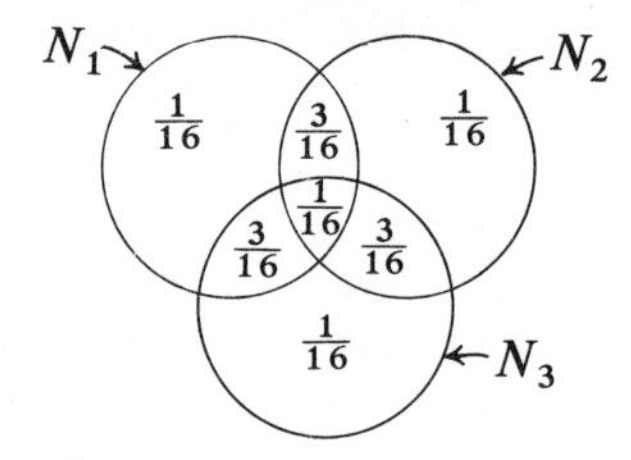


In this measure, $d$, we have

$$d(N_1) = d(N_2) = d(N_3) = \tfrac{1}{16} + \tfrac{3}{16} + \tfrac{3}{16} + \tfrac{1}{16} = \tfrac{1}{2}$$
$$d(N_1 \cap N_2) = d(N_1 \cap N_3) = d(N_2 \cap N_3) = \tfrac{3}{16} + \tfrac{1}{16} = \tfrac{1}{4}$$
$$d(N_1) \cdot d(N_2) = d(N_1) \cdot d(N_3) = d(N_2) \cdot d(N_3) = \tfrac{1}{4}$$
$$d(N_1 \cap N_2 \cap N_3) = \tfrac{1}{16} \neq d(N_1) \cdot d(N_2 \cap N_3) = \tfrac{1}{8}.$$

What is needed to assure the complete independence of each event $N_i$ from every collection of others is the following constraint:

Let $\mathfrak{L}$ be any collection of the events $N_i$ or any constitution of such a set. Then if neither $N_k$ nor $-N_k$ is a member of $\mathfrak{L}$, $N_k$ is independent of $\bigcap \mathfrak{L}$ in $c$, or, equivalently;

Let $\mathfrak{L}$ be any collection of the events $N_i$ or any constitution of such a set. Then $\mathfrak{L}$ is independent in $c$, i.e.,

$$c(\bigcap \mathfrak{L}) = \prod_{X \in \mathfrak{L}} c(X).$$

This, in addition to the requirement

$$c(N_i) = \tfrac{1}{2}, \quad \text{for each } i$$

is sufficient to determine completely the measure $c$.

We have some easy and interesting consequences of these constraints.

(1) If $\mathfrak{R}$ is a finite collection of the events $N_i$, or a constitution of such a collection, then $c(\bigcap \mathfrak{R}) = \frac{1}{2}^k$ where $k$ is the size of $\mathfrak{R}$.

(2) If $\mathfrak{R}$ is an infinite collection of the events $N_i$, or a constitution of such a collection, then $c(\bigcap \mathfrak{R}) = 0$.

*Proof* of (2). Let $\mathfrak{R} = \{X_1, X_2, \ldots\}$ where each $X_j$ is either an event $N_i$ or the complement of such an event, and such that if $N_i$ is in $\mathfrak{R}$ then $-N_i$ is not in $\mathfrak{R}$.

Let

$$\begin{aligned} \mathfrak{R}_0 &= \Lambda \\ \mathfrak{R}_k &= \{X_1, \ldots X_k\} \end{aligned}$$

Then

$$c(\bigcap \mathfrak{R}_k) = \tfrac{1}{2}^k$$

and

$$\begin{aligned} c(\bigcap \mathfrak{R}) &= c(\lim_{k \to \infty} \bigcap \mathfrak{R}_k) \\ &= \lim_{k \to \infty} c(\bigcap \mathfrak{R}_k) \\ &= \lim_{k \to \infty} \tfrac{1}{2}^k = 0. \end{aligned}$$

# BIBLIOGRAPHY

Bayes, Thomas, 'An Essay Toward Solving a Problem in the Doctrine of Chances', *Biometrika*, **45** (1958).

Brentano, Franz, *Psychology from an Empirical Standpoint*, (Translated by Rancurello *et al.*), New York, 1973.

Carnap, Rudolf, *Meaning and Necessity*, Chicago, 1956. Second edition.

Carnap, Rudolf, 'Psychology in Physical Language', in *Logical Positivism*, Edited by A. J. Ayer, Glencoe, Ill., 1959.

Carnap, Rudolf, 'The Aim of Inductive Logic', in *Logic, Methodology, and Philosophy of Science*, Edited by E. Nagel, P. Suppes, and A. Tarski, Stanford, 1962.

Carnap, Rudolf, *Logical Foundations of Probability*, Chicago, 1962, (Second edition).

Carnap, Rudolf and Richard Jeffrey, *Studies in Inductive Logic and Probability*, Volume I, Berkeley and Los Angeles, 1971.

Chisholm, Roderick, *Realism and the Background of Phenomenology*, Glencoe, Ill., 1960.

Chomsky, Noam, *Aspects of the Theory of Syntax*, Cambridge, 1965.

Church, Alonzo, 'Review of Carnap's *Introduction to Semantics*', *Philosophical Review*, **52** (1943), pp. 45–47.

De Finetti, Bruno, La prévision: ses lois, ses sources subjectives, *Annales de l'institut Henri Poincaré*. Volume **7** (1937).

Gaifman, Chaim, 'Concerning Measures on First-Order Calculi', *Israel Journal of Mathematics*, Volume **2** (1964), pp. 1–18.

Good, I. J., 'Subjective Probability as The Measure of a Non-Measurable Set', in *Logic, Methodology, and Philosophy of Science*, Edited by E. Nagel, P. Suppes, and A. Tarski, Stanford, 1962.

Hume, David, *A Treatise of Human Nature*, Edited by L. A. Selby-Bigge, Oxford, 1955 (Second edition).

Jeffrey, Richard, *The Logic of Decision*, New York, 1965.

Jeffrey, Richard and Carnap, Rudolf, see Carnap and Jeffrey.

Kaplan, David, 'Quantifying In', In *Words and Objections*. Edited by Donald Davidson and Jaakko Hintikka, Dordrecht, 1969.

Koopman, B. O., 'The Axioms and Algebra of Intuitive Probability', *Annals of Mathematics*, Volume **41** (1940), pp. 763–774.

Krauss, Peter and Scott, Dana, see Scott and Krauss.

Kuhn, Thomas, *The Structure of Scientific Revolutions*, Chicago, 1962.

Kyburg, Henry, and Smokler, Howard (editors), *Studies in Subjective Probability*, New York, 1964.

Marcuse, Herbert, *Negations*, Boston, 1968.

Meinong, Alexius, 'The Theory of Objects', translated by Levi *et al.*, in *Realism and the Background of Phenomenology*.

Peirce, Charles, 'The Probability of Induction', *Popular Science Monthly*, April, 1878. Also found in *Chance, Love, and Logic*, London, 1923.

Quine, Willard, 'Reference and Modality', in *From a Logical Point of View*.

Quine, Willard, *Word and Object,* New York and London, 1960.
Quine, Willard, *From a Logical Point of View,* New York, 1963, (Second edition).
Ramsey, F. P., 'General Propositions and Causality', in *Foundations of Mathematics.*
Ramsey, F. P., *Foundations of Mathematics and Other Essays,* London, 1931.
Russell, Bertrand, *The Principles of Mathematics,* London, 1903.
Russell, Bertrand, 'Meinong's Theory of Complexes and Assumptions', *Mind* 1904.
Russell, Bertrand, *Philosophical Essays,* London, 1910.
Russell, Bertrand, *Logic and Knowledge,* Edited by C. Marsh, London and New York, 1956.
Russell, Bertrand, 'On The Nature of Truth and Falsehood', in *Philosophical Essays.*
Russell, Bertrand, 'The Philosophy of Logical Atomism', in *Logic and Knowledge.*
Russell, Bertrand, 'On Propositions, What They Are and How They Mean', in *Philosophical Essays.*
Russell, Bertrand and Whitehead, Alfred, see Whitehead and Russell.
Ryle, Gilbert, *The Concept of Mind,* London, 1949.
Savage, Leonard, *The Foundations of Statistics,* New York, 1954.
Scheffler, Israel, *The Anatomy of Inquiry,* New York, 1963.
Schoenfield, Joseph, *Mathematical Logic,* New York, 1967.
Scott, Dana, and Suppes, Patrick, 'Foundational Aspects of Theories of Measurement', *Journal of Symbolic Logic,* **23** (1958), pp. 113–128.
Scott, Dana and Krauss, Peter, 'Assigning Probabilities to Logical Formulas', in *Aspects of Inductive Logic.* Edited by Jaakko Hintikka and Patrick Suppes, Amsterdam, 1966.
Shimony, Abner, 'Coherence and The Axioms of Confirmation', *Journal of Symbolic Logic,* **20** (1955), pp. 1–28.
Smokler, Howard, and Kyburg, Henry, see Kyburg and Smokler.
Suppes, Patrick, 'The Role of Subjective Probability in Decision Making', *Proceedings of The Third Berkeley Symposium on Mathematics, Statistics, and Probability,* 1954–1955. Volume **5**, pp. 6L–73.
Suppes, Patrick and Scott, Dana. See Scott and Suppes.
Suppes, Patrick and Zinnes, J., 'Basic Measurement Theory', in *Handbook of Mathematical Psychology,* Volume I, New York, 1963.
Tarski, Alfred, *Logic, Semantics, and Metamathematics,* Oxford, 1956.
Tarski, Alfred, 'The Concept of Truth in Formalized Languages', in *Logic, Semantics and Metamathematics.*
Van Fraassen, Bas, *Formal Semantics and Logic*, New York, 1971.
Vickers, J., 'Probability and Non-standard Logics', in *Some Recent Developments on Philosophical Problems in Logic,* Edited by Karel Lambert, Dordrecht, 1972.
Whitehead, Alfred and Russell, Bertrand, *Principia Mathematica,* Cambridge, 1957, (Second edition).
Zinnes, Joseph, and Suppes, Patrick, see Suppes and Zinnes.

# INDEX OF NAMES AND SUBJECTS*



* For technical terms and notation, only the definitions are cited

# SYNTHESE LIBRARY

Monographs on Epistemology, Logic, Methodology,
Philosophy of Science, Sociology of Science and of Knowledge, and on the
Mathematical Methods of Social and Behavioral Sciences



1. J. M. BOCHEŃSKI, *A Precis of Mathematical Logic*. 1959, X + 100 pp.
2. P. L. GUIRAUD, *Problèmes et méthodes de la statistique linguistique*. 1960, VI + 146 pp.
3. HANS FREUDENTHAL (ed.), *The Concept and the Role of the Model in Mathematics and Natural and Social Sciences, Proceedings of a Colloquium held at Utrecht, The Netherlands, January 1960*. 1961, VI + 194 pp.
4. EVERT W. BETH, *Formal Methods. An Introduction to Symbolic Logic and the Study of Effective Operations in Arithmetic and Logic*. 1962, XIV + 170 pp.
5. B. H. KAZEMIER and D. VUYSJE (eds.), *Logic and Language. Studies dedicated to Professor Rudolf Carnap on the Occasion of his Seventieth Birthday*. 1962, VI + 256 pp.
6. MARX W. WARTOFSKY (ed.), *Proceedings of the Boston Colloquium for the Philosophy of Science, 1961–1962*, Boston Studies in the Philosophy of Science (ed. by Robert S. Cohen and Marx W. Wartofsky), Volume I. 1963, VIII + 212 pp.
7. A. A. ZINOV'EV, *Philosophical Problems of Many-Valued Logic*. 1963, XIV + 155 pp.
8. GEORGES GURVITCH, *The Spectrum of Social Time*. 1964, XXVI + 152 pp.
9. PAUL LORENZEN, *Formal Logic*. 1965, VIII + 123 pp.
10. ROBERT S. COHEN and MARX W. WARTOFSKY (eds.), *In Honor of Philipp Frank*, Boston Studies in the Philosophy of Science (ed. by Robert S. Cohen and Marx W. Wartofsky), Volume II. 1965, XXXIV + 475 pp.
11. EVERT W. BETH, *Mathematical Thought. An Introduction to the Philosophy of Mathematics*. 1965, XII + 208 pp.
12. EVERT W. BETH and JEAN PIAGET, *Mathematical Epistemology and Psychology*. 1966, XII + 326 pp.
13. GUIDO KÜNG, *Ontology and the Logistic Analysis of Language. An Enquiry into the Contemporary Views on Universals*. 1967, XI + 210 pp.

14. ROBERT S. COHEN and MARX W. WARTOFSKY (eds.), *Proceedings of the Boston Colloquium for the Philosophy of Science 1964–1966, in Memory of Norwood Russell Hanson.* Boston Studies in the Philosophy of Science (ed. by Robert S. Cohen and Marx W. Wartofsky), Volume III. 1967, XLIX + 489 pp.
15. C. D. BROAD, *Induction, Probability, and Causation. Selected Papers.* 1968, XI + 296 pp.
16. GÜNTHER PATZIG, *Aristotle's Theory of the Syllogism. A Logical-Philosophical Study of Book A of the Prior Analytics.* 1968, XVII + 215 pp.
17. NICHOLAS RESCHER, *Topics in Philosophical Logic.* 1968, XIV + 347 pp.
18. ROBERT S. COHEN and MARX W. WARTOFSKY (eds.), *Proceedings of the Boston Colloquium for the Philosophy of Science 1966–1968*, Boston Studies in the Philosophy of Science (ed. by Robert S. Cohen and Marx W. Wartofsky), Volume IV. 1969, VIII + 537 pp.
19. ROBERT S. COHEN and MARX W. WARTOFSKY (eds.), *Proceedings of the Boston Colloquium for the Philosophy of Science 1966–1968,* Boston Studies in the Philosophy of Science (ed. by Robert S. Cohen and Marx W. Wartofsky), Volume V. 1969, VIII + 482 pp.
20. J. W. DAVIS, D. J. HOCKNEY, and W. K. WILSON (eds.), *Philosophical Logic.* 1969, VIII + 277 pp.
21. D. DAVIDSON and J. HINTIKKA (eds.), *Words and Objections: Essays on the Work of W. V. Quine.* 1969, VIII + 366 pp.
22. PATRICK SUPPES, *Studies in the Methodology and Foundations of Science. Selected Papers from 1911 to 1969,* XII + 473 pp.
23. JAAKKO HINTIKKA, *Models for Modalities. Selected Essays.* 1969, IX + 220 pp.
24. NICHOLAS RESCHER *et al.* (eds.), *Essays in Honor of Carl G. Hempel. A Tribute on the Occasion of his Sixty-Fifth Birthday.* 1969, VII + 272 pp.
25. P. V. TAVANEC (ed.), *Problems of the Logic of Scientific Knowledge.* 1969, XII + 429 pp.
26. MARSHALL SWAIN (ed.), *Induction, Acceptance, and Rational Belief.* 1970, VII + 232 pp.
27. ROBERT S. COHEN and RAYMOND J. SEEGER (eds.), *Ernst Mach; Physicist and Philosopher,* Boston Studies in the Philosophy of Science (ed. by Robert S. Cohen and Marx W. Wartofsky), Volume VI. 1970, VIII + 295 pp.
28. JAAKKO HINTIKKA and PATRICK SUPPES, *Information and Inference.* 1970, X + 336 pp.
29. KAREL LAMBERT, *Philosophical Problems in Logic. Some Recent Developments.* 1970, VII + 176 pp.
30. ROLF A. EBERLE, *Nominalistic Systems.* 1970, IX + 217 pp.
31. PAUL WEINGARTNER and GERHARD ZECHA (eds.), *Induction, Physics, and Ethics, Proceedings and Discussions of the 1968 Salzburg Colloquium in the Philosophy of Science.* 1970, X + 382 pp.
32. EVERT W. BETH, *Aspects of Modern Logic.* 1970, XI + 176 pp.
33. RISTO HILPINEN (ed.), *Deontic Logic: Introductory and Systematic Readings.* 1971, VII + 182 pp.
34. JEAN-LOUIS KRIVINE, *Introduction to Axiomatic Set Theory.* 1971, VII + 98 pp.
35. JOSEPH D. SNEED, *The Logical Structure of Mathematical Physics.* 1971, XV + 311 pp.
36. CARL R. KORDIG, *The Justification of Scientific Change.* 1971, XIV + 119 pp.
37. MILIČ ČAPEK, *Bergson and Modern Physics,* Boston Studies in the Philosophy of Science (ed. by Robert S. Cohen and Marx W. Wartofsky), Volume VII. 1971, XV + 414 pp.
38. NORWOOD RUSSELL HANSON, *What I do not Believe, and other Essays,* (ed. by Stephen Toulmin and Harry Woolf), 1971, XII + 390 pp.
39. ROGER C. BUCK and ROBERT S. COHEN (eds.), *PSA 1970. In Memory of Rudolf Carnap,* Boston Studies in the Philosophy of Science (ed. by Robert S. Cohen and Marx W. Wartofsky), Volume VIII. 1971, LXVI + 615 pp. Also available as a paperback.

40. DONALD DAVIDSON and GILBERT HARMAN (eds.), *Semantics of Natural Language*. 1972, X+769 pp. Also available as a paperback.
41. YEHOSHUA BAR-HILLEL (ed.), *Pragmatics of Natural Languages*. 1971, VII+231 pp.
42. SÖREN STENLUND, *Combinators, λ-Terms and Proof Theory*. 1972, 184 pp.
43. MARTIN STRAUSS, *Modern Physics and Its Philosophy. Selected Papers in the Logic. History, and Philosophy of Science*. 1972, X+297 pp.
44. MARIO BUNGE, *Method, Model and Matter*. 1973, VII+196 pp.
45. MARIO BUNGE, *Philosophy of Physics*. 1973, IX+248 pp.
46. A. A. ZINOV'EV, *Foundations of the Logical Theory of Scientific Knowledge (Complex Logic)*, Boston Studies in the Philosophy of Science (ed. by Robert S. Cohen and Marx W. Wartofsky), Volume IX. Revised and enlarged English edition with an appendix, by G. A. Smirnov, E. A. Sidorenka, A. M. Fedina, and L. A. Bobrova 1973, XXII+301 pp. Also available as a paperback.
47. LADISLAV TONDL, *Scientific Procedures*, Boston Studies in the Philosophy of Science (ed. by Robert S. Cohen and Marx W. Wartofsky), Volume X. 1973, XII+268 pp. Also available as a paperback.
48. NORWOOD RUSSELL HANSON, *Constellations and Conjectures*, (ed. by Willard C. Humphreys, Jr.), 1973, X+282 pp.
49. K. J. J. HINTIKKA, J. M. E. MORAVCSIK, and P. SUPPES (eds.), *Approaches to Natural Language. Proceedings of the 1970 Stanford Workshop on Grammar and Semantics*. 1973, VIII+526 pp. Also available as a paperback.
50. MARIO BUNGE (ed.), *Exact Philosophy – Problems, Tools, and Goals*. 1973, X+214 pp.
51. RADU J. BOGDAN and ILKKA NIINILUOTO (eds.), *Logic, Language, and Probability*. A selection of papers contributed to Sections IV, VI, and XI of the Fourth International Congress for Logic, Methodology, and Philosophy of Science, Bucharest, September 1971. 1973, X+323 pp.
52. GLENN PEARCE and PATRICK MAYNARD (eds.), *Conceptual Chance*. 1973, XII+282 pp.
53. ILKKA NIINILUOTO and RAIMO TUOMELA, *Theoretical Concepts and Hypothetico-Inductive Inference*. 1973, VII+264 pp.
54. ROLAND FRAÏSSÉ, *Course of Mathematical Logic* – Volume 1: *Relation and Logical Formula*. 1973, XVI+186 pp. Also available as a paperback.
55. ADOLF GRÜNBAUM, *Philosophical Problems of Space and Time*. Second, enlarged edition, Boston Studies in the Philosophy of Science (ed. by Robert S. Cohen and Marx W. Wartofsky), Volume XII. 1973, XXIII+884 pp. Also available as a paperback.
56. PATRICK SUPPES (ed.), *Space, Time, and Geometry*. 1973, XI+424 pp.
57. HANS KELSEN, *Essays in Legal and Moral Philosophy*, selected and introduced by Ota Weinberger. 1973, XXVIII+300 pp.
58. R. J. SEEGER and ROBERT S. COHEN (eds.), *Philosophical Foundations of Science. Proceedings of an AAAS Program, 1969*. Boston Studies in the Philosophy of Science (ed. by Robert S. Cohen and Marx W. Wartofsky), Volume XI. 1974, X+545 pp. Also available as a paperback.
59. ROBERT S. COHEN and MARX W. WARTOFSKY (eds.), *Logical and Epistemological Studies in Contemporary Physics*, Boston Studies in the Philosophy of Science (ed. by Robert S. Cohen and Marx W. Wartofsky), Volume XIII. 1973, VIII+462 pp. Also available as paperback.
60. ROBERT S. COHEN and MARX WARTOFSKY (eds.), *Methodological and Historical Essays in the Natural and Social Sciences. Proceedings of the Boston Colloquium for the Philosophy of Science, 1969–1972*, Boston Studies in the Philosophy of Science (ed. by Robert S. Cohen and Marx W. Wartofsky), Volume XIV. 1974, VIII+405 pp. Also available as paperback.
61. ROBERT S. COHEN, J. J. STACHEL and MARX W. WARTOFSKY (eds.), *For Dirk Struik*

*Scientific, Historical and Political Essays in Honor of Dirk J. Struik*, Boston Studies in the Philosophy of Science (ed. by Robert S. Cohen and Marx W. Wartofsky), Volume XV. 1974, XXVII + 652 pp. Also available as paperback.

62. Kazimierz Ajdukiewicz, *Pragmatic Logic*, transl. from the Polish by Olgierd Wojtasiewicz. 1974, XV + 460 pp.
63. Sören Stenlund (ed.), *Logical Theory and Semantic Analysis. Essays Dedicated to Stig Kanger on His Fiftieth Birthday*. 1974, V + 217 pp.
64. Kenneth F. Schaffner and Robert S. Cohen (eds.), *Proceedings of the 1972 Biennial Meeting, Philosophy of Science Association*, Boston Studies in the Philosophy of Science (ed. by Robert S. Cohen and Marx W. Wartofsky), Volume XX. 1974, IX + 444 pp. Also available as paperback.
65. Henry E. Kyburg, Jr., *The Logical Foundations of Statistical Inference*. 1974, IX + 421 pp.
66. Marjorie Grene, *The Understanding of Nature: Essays in the Philosophy of Biology*, Boston Studies in the Philosophy of Science (ed. by Robert S. Cohen and Marx W. Wartofsky), Volume XXIII. 1974, XII + 360 pp. Also available as paperback.
67. Jan M. Broekman, *Structuralism: Moscow, Prague, Paris*. 1974, IX + 117 pp.
68. Norman Geschwind, *Selected Papers on Language and the Brain*, Boston Studies in the Philosophy of Science (ed. by Robert S. Cohen and Marx W. Wartofsky), Volume XVI. 1974, XII + 549 pp. Also available as paperback.
69. Roland Fraïssé, *Course of Mathematical Logic* – Volume II: *Model Theory*. 1974, XIX + 192 pp.
70. Andrzej Grzegorczyk, *An Outline of Mathematical Logic. Fundamental Results and Notions Explained with All Details*. 1974, X + 596 pp.
71. Franz von Kutschera, *Philosophy of Language*. 1975, VII + 305 pp.
72. Juha Manninen and Raimo Tuomela (eds.), *Essays on Explanation and Understanding. Studies in the Foundations of Humanities and Social Sciences*. 1976, VII + 440 pp.
73. Jaakko Hintikka (ed.), *Rudolf Carnap, Logical Empiricist. Materials and Perspectives*. 1975, LXVIII + 400 pp.
74. Milič Čapek (ed.), *The Concepts of Space and Time. Their Structure and Their Development*. Boston Studies in the Philosophy of Science (ed. by Robert S. Cohen and Marx W. Wartofsky), Volume XXII. 1976, LVI + 570 pp. Also available as paperback.
75. Jaakko Hintikka and Unto Remes, *The Method of Analysis. Its Geometrical Origin and Its General Significance*. Boston Studies in the Philosophy of Science (ed. by Robert S. Cohen and Marx W. Wartofsky), Volume XXV. 1974, XVIII + 144 pp. Also available as paperback.
76. John Emery Murdoch and Edith Dudley Sylla, *The Cultural Context of Medieval Learning. Proceedings of the First International Colloquium on Philosophy, Science, and Theology in the Middle Ages – September 1973*. Boston Studies in the Philosophy of Science (ed. by Robert S. Cohen and Marx W. Wartofsky), Volume XXVI. 1975, X + 566 pp. Also available as paperback.
77. Stefan Amsterdamski, *Between Experience and Metaphysics. Philosophical Problems of the Evolution of Science*. Boston Studies in the Philosophy of Science (ed. by Robert S. Cohen and Marx W. Wartofsky), Volume XXXV. 1975, XVIII + 193 pp. Also available as paperback.
78. Patrick Suppes (ed.), *Logic and Probability in Quantum Mechanics*. 1976, XV + 541 pp.
80. Joseph Agassi, *Science in Flux*. Boston Studies in the Philosophy of Science (ed. by Robert S. Cohen and Marx W. Wartofsky), Volume XXVIII. 1975, XXVI + 553 pp. Also available as paperback.
81. Sandra G. Harding (ed.), *Can Theories Be Refuted? Essays on the Duhem-Quine Thesis*. 1976, XXI + 318 pp. Also available in paperback.

84. Marjorie Grene and Everett Mendelsohn (eds.), *Topics in the Philosophy of Biology.* Boston Studies in the Philosophy of Science (ed. by Robert S. Cohen and Marx W. Wartofsky), Volume XXVII. 1976, XIII+454 pp. Also available as paperback.
85. E. Fischbein, *The Intuitive Sources of Probabilistic Thinking in Children.* 1975, XIII+204 pp.
86. Ernest W. Adams, *The Logic of Conditionals. An Application of Probability to Deductive Logic.* 1975, XIII+156 pp.
89. A. Kasher (ed.), *Language in Focus: Foundations, Methods and Systems. Essays dedicated to Yehoshua Bar-Hillel.* Boston Studies in the Philosophy of Science (ed. by Robert S. Cohen and Marx W. Wartofsky), Volume XLIII. 1976, XXVIII+679 pp. Also available as paperback.
90. Jaakko Hintikka, *The Intentions of Intentionality and Other New Models for Modalities.* 1975, XVIII+262 pp. Also available as paperback.
93. Radu J. Bogdan, *Local Induction.* 1976, XIV+340 pp.
95. Peter Mittelstaedt, *Philosophical Problems of Modern Physics.* Boston Studies in the Philosophy of Science (ed. by Robert S. Cohen and Marx W. Wartofsky), Volume XVIII. 1976, X+211 pp. Also available as a paperback.
96. Gerald Holton and William Blanpied (eds.), *Science and Its Public: The Changing Relationship.* Boston Studies in the Philosophy of Science (ed. by Robert S. Cohen and Marx W. Wartofsky), Volume XXXIII. 1976, XXV+289 pp. Also available as paperback.

# SYNTHESE HISTORICAL LIBRARY

Texts and Studies
in the History of Logic and Philosophy

*Editors:*



1. M. T. BEONIO-BROCCHIERI FUMAGALLI, *The Logic of Abelard.* Translated from the Italian. 1969, IX + 101 pp.
2. GOTTFRIED WILHELM LEIBNITZ, *Philosophical Papers and Letters.* A selection translated and edited, with an introduction, by Leroy E. Loemker. 1969, XII + 736 pp.
3. ERNST MALLY, *Logische Schriften,* ed. by Karl Wolf and Paul Weingartner. 1971, X + 340 pp.
4. LEWIS WHITE BECK (ed.), *Proceedings of the Third International Kant Congress.* 1972, XI + 718 pp.
5. BERNARD BOLZANO, *Theory of Science,* ed. by Jan Berg. 1973, XV + 398 pp.
6. J. M. E. MORAVCSIK (ed.), *Patterns in Plato's Thought. Papers arising out of the 1971 West Coast Greek Philosophy Conference.* 1973, VIII + 212 pp.
7. NABIL SHEHABY, *The Propositional Logic of Avicenna: A Translation from al-Shifā: al-Qiyās,* with Introduction, Commentary and Glossary. 1973, XIII + 296 pp.
8. DESMOND PAUL HENRY, *Commentary on De Grammatico: The Historical-Logical Dimensions of a Dialogue of St. Anselm's.* 1974, IX + 345 pp.
9. JOHN CORCORAN, *Ancient Logic and Its Modern Interpretations.* 1974, X + 208 pp.
10. E. M. BARTH, *The Logic of the Articles in Traditional Philosophy.* 1974, XXVII + 533 pp.
11. JAAKKO HINTIKKA, *Knowledge and the Known. Historical Perspectives in Epistemology.* 1974, XII + 243 pp.
12. E. J. ASHWORTH, *Language and Logic in the Post-Medieval Period.* 1974, XIII + 304 pp.
13. ARISTOTLE, *The Nicomachaen Ethics.* Translated with Commentaries and Glossary by Hypocrates G. Apostle. 1975, XXI + 372 pp.
14. R. M. DANCY, *Sense and Contradiction: A Study in Aristotle.* 1975, XII + 184 pp.
15. WILBUR RICHARD KNORR, *The Evolution of the Euclidean Elements. A Study of the Theory of Incommensurable Magnitudes and Its Significance for Early Greek Geometry.* 1975, IX + 374 pp.
16. AUGUSTINE, *De Dialectica.* Translated with the Introduction and Notes by B. Darrell Jackson. 1975, XI + 151 pp.